Hyperbolic and Viscous Conservation Laws

CBMS-NSF REGIONAL CONFERENCE SERIES
IN APPLIED MATHEMATICS

A series of lectures on topics of current research interest in applied mathematics under the direction of the Conference Board of the Mathematical Sciences, supported by the National Science Foundation and published by SIAM.

GARRETT BIRKHOFF, *The Numerical Solution of Elliptic Equations*

D. V. LINDLEY, *Bayesian Statistics, A Review*

R. S. VARGA, *Functional Analysis and Approximation Theory in Numerical Analysis*

R. R. BAHADUR, *Some Limit Theorems in Statistics*

PATRICK BILLINGSLEY, *Weak Convergence of Measures: Applications in Probability*

J. L. LIONS, *Some Aspects of the Optimal Control of Distributed Parameter Systems*

ROGER PENROSE, *Techniques of Differential Topology in Relativity*

HERMAN CHERNOFF, *Sequential Analysis and Optimal Design*

J. DURBIN, *Distribution Theory for Tests Based on the Sample Distribution Function*

SOL I. RUBINOW, *Mathematical Problems in the Biological Sciences*

P. D. LAX, *Hyperbolic Systems of Conservation Laws and the Mathematical Theory of Shock Waves*

I. J. SCHOENBERG, *Cardinal Spline Interpolation*

IVAN SINGER, *The Theory of Best Approximation and Functional Analysis*

WERNER C. RHEINBOLDT, *Methods of Solving Systems of Nonlinear Equations*

HANS F. WEINBERGER, *Variational Methods for Eigenvalue Approximation*

R. TYRRELL ROCKAFELLAR, *Conjugate Duality and Optimization*

SIR JAMES LIGHTHILL, *Mathematical Biofluiddynamics*

GERARD SALTON, *Theory of Indexing*

CATHLEEN S. MORAWETZ, *Notes on Time Decay and Scattering for Some Hyperbolic Problems*

F. HOPPENSTEADT, *Mathematical Theories of Populations: Demographics, Genetics and Epidemics*

RICHARD ASKEY, *Orthogonal Polynomials and Special Functions*

L. E. PAYNE, *Improperly Posed Problems in Partial Differential Equations*

S. ROSEN, *Lectures on the Measurement and Evaluation of the Performance of Computing Systems*

HERBERT B. KELLER, *Numerical Solution of Two Point Boundary Value Problems*

J. P. LASALLE, *The Stability of Dynamical Systems* - Z. ARTSTEIN, *Appendix A: Limiting Equations and Stability of Nonautonomous Ordinary Differential Equations*

D. GOTTLIEB AND S. A. ORSZAG, *Numerical Analysis of Spectral Methods: Theory and Applications*

PETER J. HUBER, *Robust Statistical Procedures*

HERBERT SOLOMON, *Geometric Probability*

FRED S. ROBERTS, *Graph Theory and Its Applications to Problems of Society*

JURIS HARTMANIS, *Feasible Computations and Provable Complexity Properties*

ZOHAR MANNA, *Lectures on the Logic of Computer Programming*

ELLIS L. JOHNSON, *Integer Programming: Facets, Subadditivity, and Duality for Group and Semi-Group Problems*

SHMUEL WINOGRAD, *Arithmetic Complexity of Computations*

J. F. C. KINGMAN, *Mathematics of Genetic Diversity*

MORTON E. GURTIN, *Topics in Finite Elasticity*

THOMAS G. KURTZ, *Approximation of Population Processes*

Tai-Ping Liu
Stanford University
Stanford, California

Hyperbolic and Viscous Conservation Laws

SOCIETY FOR INDUSTRIAL AND APPLIED MATHEMATICS

PHILADELPHIA

Library of Congress Cataloging-in-Publication Data

Liu, Tai-Ping, 1945-
    Hyperbolic and viscous conservation laws / Tai-Ping Liu.
        p. cm. -- (CBMS-NSF regional conference series in applied mathematics ; 72)
    Includes bibliographical references and index.
    ISBN 0-89871-436-2 (pbk.)
    1. Conservation laws (Mathematics) 2. Shock waves--Mathematics. I. Title.
    II. Series.

QA377.L568 2000
531'.1133--dc21

                                                                    99-049815

siam is a registered trademark.

Contents

Preface

There has been intense interest in recent years in the study of shock waves in various physical situations: nonlinear elasticity, magnetohydrodynamics, multiphase flows, combustion, etc., in addition to the classical gas dynamics shocks. The purpose of this study is to understand, physically and mathematically, the new wave phenomena. These issues are very interesting because of the strongly nonlinear effects. The central issue is the understanding of nonlinear wave interactions. I have presented in this book, for the first time, the in-depth analysis of wave interactions for general systems of hyperbolic and viscous conservation laws.

The book starts with the basic ideas of shock wave theory and is suitable for graduate students interested in an introduction to this vital area of nonlinear analysis. The book is also aimed at researchers who are interested in nonlinear waves in general and would like to become familiarized with the analytical techniques that have been introduced, some in the last few years, for the qualitative theory of shock waves.

I will discuss the existence, regularity, and large-time behavior of solutions for hyperbolic conservation laws. The main tools are the random choice method and the wave tracing technique. For viscous conservation laws, I have included the recent analysis of dissipation, compression, and expansion waves. Energy and pointwise estimates are the main analytical techniques. The main theme of these two studies is the treatment of nonlinear interaction of waves. I will start with the scalar conservation law to illustrate the elementary notions of weak solutions and entropy conditions. For viscous conservation laws, I will start with the Burgers equation to understand the coupling of nonlinear flux and dissipation.

ACKNOWLEDGMENTS

This book is an outgrowth of the CBMS Regional Lectures held at Georgia Tech, June 1997. The author wishes to thank Professors Shui-Nee Chow and Shi Jin for organizing and hosting the lectures.

The research of the author is supported in part by NSF grant DMS-9803323.

Tai-Ping Liu

Hyperbolic Conservation Laws

1.1 Preliminaries

Consider the system of conservation laws

(1.1.1) $$\mathbf{u}_t + f(\mathbf{u})_x = 0.$$

Here $\mathbf{u} = \mathbf{u}(x,t)$ and $f(\mathbf{u})$ are n-vectors representing, respectively, the density of some physical quantities and the flux function. This is the simplest model of balance laws, assuming that the flux $f(\mathbf{u}(x,t))$ depends only on the local value of the physical density $\mathbf{u}(x,t)$ and there is no sink or source. The time variable t is nonnegative and the space variable x is taken to be in $\mathbb{R}^1$. We assume that the system carries infinitesimal waves or, equivalently, is *hyperbolic*; that is, the eigenvalues of the $n \times n$ matrix $f'(\mathbf{u})$ are real:

$$f'(\mathbf{u})r_i(\mathbf{u}) = \lambda_i(\mathbf{u})r_i(\mathbf{u}),$$
$$l_i(\mathbf{u})f'(\mathbf{u}) = \lambda_i(\mathbf{u})l_i(\mathbf{u}),$$
$$i = 1, 2, \ldots, n,$$

(1.1 2) $$\lambda_1(\mathbf{u}) \leq \lambda_2(\mathbf{u}) \leq \cdots \leq \lambda_n(\mathbf{u}).$$

System (1.1.1) is *completely hyperbolic* if the eigenvectors $r_i(\mathbf{u}), i = 1, 2, \ldots, n$, form a basis in $\mathbb{R}^n$, and it is *strictly hyperbolic* if the eigenvalues, the *characteristics*, are distinct: $\lambda_1(\mathbf{u}) < \lambda_2(\mathbf{u}) < \cdots < \lambda_n(\mathbf{u})$. From linear algebra we know that strict hyperbolicity implies complete hyperbolicity, and, in either case, the eigenvectors can be normalized as

(1.1.3) $$r_i(\mathbf{u}) \cdot l_j(\mathbf{u}) = \delta_{ij}, \qquad i, j = 1, 2, \ldots, n.$$

An important physical example is the Euler equations for gas dynamics [2]:

$$\rho_t + (\rho v)_x = 0,$$
$$(\rho v)_t + (\rho v^2 + p)_x = 0,$$
(1.1.4) $$(\rho E)_t + (\rho E v + p v)_x = 0.$$

These represent the conservation of mass, momentum, and energy. It is assumed for the Euler equations that the only internal force is the pressure, which is taken to be a known function of density ρ and internal energy e through the equation of state: $p = p(\rho, e)$. The equation of state defines the gas. The velocity of the gas is v and the total energy $E = v^2/2 + e$. The eigenvalues are v, $v \pm c$, where c is the sound speed, given by $c^2 = pp_e + \rho p_\rho$. At least for the polytropic gases, $p = (\gamma - 1)\rho e$, $1 < \gamma \le 5/3$, the sound speed is real and nonzero when the gas is not in vacuum, $\rho > 0$. Other physical systems such as the magnetohydrodynamics and full nonlinear elasticity equations are completely hyperbolic but not strictly hyperbolic. In this chapter we assume that (1.1.1) is strictly hyperbolic. Because of the dependence of the characteristics $\lambda_i(\mathbf{u})$ on the dependent variables $\mathbf{u}$, waves may compress to form *shock waves*, and in general smooth solutions do not exist globally in time. One is forced to consider the *weak solutions* [25].

DEFINITION 1.1.1. *A bounded measurable function* $\mathbf{u}(x,t)$ *is a weak solution of* (1.1.1) *with given initial value* $\mathbf{u}(x,0)$ *if and only if*

$$(1.1.5) \qquad \int_0^\infty \int_{-\infty}^\infty [\phi_t \mathbf{u} + \phi_x f(\mathbf{u})](x,t)dxdt + \int_{-\infty}^\infty (\phi \mathbf{u})(x,0)dx = 0$$

for any smooth function $\phi(x,t)$ *with compact support in* $\{(x,t)|t \ge 0,\ x \in \mathbb{R}^1\}$.

A discontinuity in the weak solution satisfies the following (*Rankine–Hugoniot*) jump condition:

$$(1.1.6) \qquad s(\mathbf{u}_+ - \mathbf{u}_-) = f(\mathbf{u}_+) - f(\mathbf{u}_-).$$

Here $s = x'(t)$ is the speed of the discontinuity $x = x(t)$ and $\mathbf{u}_\pm \equiv \mathbf{u}(x(t) \pm 0, t)$ are the states at the jump. This follows from (1.1.5) by choosing the test function to concentrate at the discontinuity. There are two basic elements in the shock theory: the concept of an entropy condition for the shock waves and the compactness properties resulting from the nonlinearity of the flux function $f(\mathbf{u})$. This compactness has important consequences on the behavior of the solutions.

Chapter 1 studies these topics, particularly the regularity and large-time behavior of the solutions. We now illustrate some of the basic issues and analytical ideas for the scalar *convex* equation

$$u \in \mathbb{R}^1, \qquad f''(u) > 0.$$

The first issue is that a weak solution with a given initial value $u(x,0)$ may not be unique. Take the example of the *Riemann problem* (1.1) with initial data

$$(1.1.7) \qquad u(x,0) = \begin{cases} u_l, & x < 0, \\ u_r, & x > 0. \end{cases}$$

In this case the characteristics are diverging $f'(u_r) > f'(u_l)$; equivalently, due to the convexity condition $f''(u) > 0$, $u_r > u_l$, there is the continuous solution, the *rarefaction wave* (see Figure 1.1.1)

$$(1.1.8) \qquad f'(u(x,t)) = \begin{cases} f'(u_l), & x < f'(u_l)t, \\ \frac{x}{t}, & f'(u_l)t < x < f'(u_r)t, \\ f'(u_r), & x > f'(u_r)t. \end{cases}$$

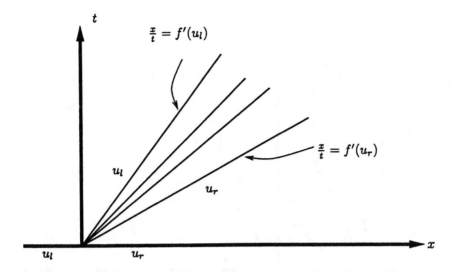

Figure 1.1.1. Rarefaction wave.

On the other hand, there is also the simple discontinuity solution $(u_-, u_+) = (u_l, u_r)$:

$$u(x,t) = \begin{cases} u_-, & x < st, \\ u_+, & x > st, \; s = \frac{f(u_+)-f(u_-)}{u_+-u_-}. \end{cases}$$

It is natural to favor the continuous solution and disallow the second solution, the rarefaction shock with the characteristics leaving the discontinuity, for which

$$f'(u_-) < s < f'(u_+).$$

The more important reason for requiring the characteristics to impinge on the shock is that the information may vanish into the shock but may not be created at a shock. This is the *Lax entropy condition* for a convex flux equation, $f''(u) \neq 0$ (see Figure 1.1.2):

(1.1.9) $$f'(u_+) < s < f'(u_-).$$

A weak solution is called *admissible* if across each discontinuity the entropy condition is satisfied. If so, the Riemann problem is solved uniquely by a rarefaction wave if $u_l < u_r$ and a shock wave if $u_l > u_r$.

We next consider the compactness property of the nonlinearity $f''(u) \neq 0$. The entropy condition says that the characteristics impinge on the shock and therefore information may be lost into the shock. For instance, two shocks next to each other will combine into a single shock; see Figure 1.1.3.

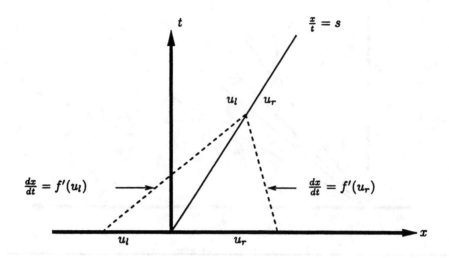

Figure 1.1.2. Shock wave.

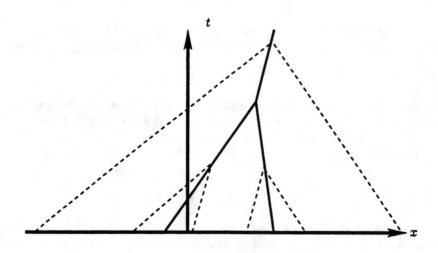

Figure 1.1.3. Shocks combining.

t

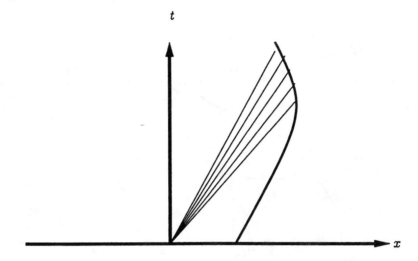

Figure 1.1.4. Cancelation of shock and rarefaction.

As another example, put a shock next to a rarefaction wave and they will cancel and eventually yield either a single shock or a rarefaction wave, whichever is stronger; see Figure 1.1.4.

Thus the process is *simplifying* and *irreversible*.

An effective technique makes use of Glimm's notion of *generalized characteristics* [12]. A generalized characteristic curve follows a characteristic line and then a shock curve upon hitting it. Thus characteristics may vanish into a generalized characteristic but not cross it. This implies, for instance, that if the solution is constant on one side of a generalized characteristic at a given time, then it remains so at a later time. In the backward time direction, a characteristic line never hits a shock as a consequence of the entropy condition. (It is important to note that this is so only for convex conservation laws $f''(u) \neq 0$. In the nonconvex case, the characteristic may be tangent to a discontinuity, the contact discontinuity, and more complex wave behavior occurs.) We now use this idea to study the stability of shock waves. Consider an initial value satisfying

$$u(x,0) = \begin{cases} u_l, & x < -M, \\ u_r, & x > M, \end{cases}$$

and suppose that $u_l > u_r$ so that the end states give rise to a shock. We will show that, for any initial value in $|x| < M$, the solution tends to a single shock,

$$u(x,t) = \begin{cases} u_l, & x - x_0 < st, \\ u_r, & x - x_0 > st, \end{cases}$$

after finite time $t > T$. The phase shift x_0 is determined easily from the conservation law that the integral of the difference between the general solution and the

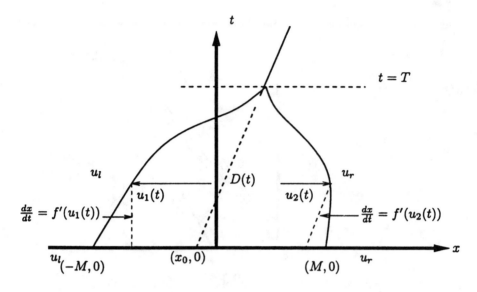

Figure 1.1.5. Stability of shock.

single shock is time invariant:

$$\frac{d}{dt}\left[\int_{-\infty}^{x_0}(u_- - u(x,t))dx + \int_{x_0}^{\infty}(u_+ - u(x,t))dx\right] = 0.$$

After time $t = T$ the solution is the single shock and the integral is zero. Thus evaluating the integral at time zero and noting that the initial data is assumed constant for $|x| > M$, we obtain

$$\frac{d}{dt}\int_{-M}^{x_0}(u_- - u(x,0))dx + \int_{x_0}^{M}(u_+ - u(x,0))dx = 0.$$

This determines the phase shift x_0. We now show that the solution tends to a single shock in finite time. Draw generalized characteristics $x = x_l(t)$ and $x = x_r(t)$ through $x = -M$ and $x = M$ at time $t = 0$, respectively. Outside these curves the solution takes the values of u_l and u_r. Thus it suffices to show that these curves meet in finite time T, $D(t) \equiv x_r(t) - x_l(t) = 0$ for $t > T$; see Figure 1.1.5.

The curves propagate with shock or characteristic speeds

$$\frac{d}{dt}D(t) = \frac{f(u_r) - f(u_2(t))}{u_r - u_2(t)} - \frac{f(u_1(t)) - f(u_l)}{u_1(t) - u_l},$$

$$u_1(t) \equiv u(x_l(t) + 0, t), \qquad u_2(t) \equiv u(x_r(t) - 0, t).$$

This follows from the jump condition. In the case where the curves are not shocks but are characteristic lines, the above quotients are understood as $f'(u)$. We now study the evolution of the distance $D(t)$. First draw the backward characteristic lines through $(x_l(t) + 0, t)$ and $(x_r(t) - 0, t)$. These have speeds $f'(u_1(t))$ and $f'(u_2(t))$ and lie between the generalized characteristics $x = x_l(t)$ and $x = x_r(t)$. Therefore, they intersect the initial time $t = 0$ between $x = -M$ and $x = M$. Thus we have

$$D(t) = (f'(u_2(t)) - f'(u_1(t)))t + O(1)$$

for a nonnegative function $O(1)$ less than $2M$. On the other hand, from the convexity of $f(u)$ and the entropy condition $u_1(t) \leq u_l$, $u_r \leq u_2(t)$, we have

$$D'(t) = \theta(t)(f'(u_2(t)) - f'(u_1(t))) + (1 - \theta(t))(f'(u_r) - f'(u_l))$$

for some function $\theta(t)$. The mean value theorem is applied here for the function $f(u)$ and the dependence on time t is only through $u = u(x, t)$. Thus $\theta(t)$ lies strictly between 0 and 1 because all the states u are in a bounded (compact) set given by the initial data. Taking $\theta \equiv \max \theta(t)$ we have from the above identities that

$$\begin{aligned} D'(t) &\leq \theta(f'(u_2(t)) - f'(u_1(t))) + (1 - \theta)(f'(u_r) - f'(u_l)) \\ &\leq \frac{\theta D(t)}{t} + (1 - \theta)(f'(u_r) - f'(u_l)). \end{aligned}$$

This is a differential inequality, which can be solved to yield

$$D(t) \leq [D(1) + (t^{1-\theta} - 1)(f'(u_r) - f'(u_l))]t^\theta.$$

Since we have assumed that the states u_l, u_r give rise to a shock, $u_l > u_r$, $f'(u_l) > f'(u_r)$, and so, by $0 < \theta < 1$, the above estimate yields

$$D(t) \equiv 0, \qquad t \geq T,$$

for some finite time T. We have thus shown that the solution consists of a single shock after finite time. In particular, the shock is nonlinearly stable and the process is highly irreversible. The stability of rarefaction waves is shown in a similar way.

The compactness property is also seen clearly in the following basic estimate on the expansion waves. Given a solution $u(x, t)$ and an interval $I(t)$ at time t with length $|I(t)|$, we have

(1.1.10) $\qquad \begin{aligned} & X_+(I(t)) \leq \frac{|I(t)|}{t}, \\ & X_+(I(t)) \equiv \text{increasing variation}\{f'(u)(x, t) : x \in I(t)\}. \end{aligned}$

The proof is simple: draw the backward characteristic lines from end points x_-, x_+ of any subinterval I' of I to yield

$$(f'(u(x_+, t)) - f'(u(x_-, t)))t \leq |I'|$$

and then sum up all the subintervals along which $f'(u)(\cdot, t)$ is increasing. An immediate consequence is that if the solution is bounded, $|f'(u(x,t))| \leq M$, then the decreasing variation $X_-(I(t))$ is no greater than $X_+(I(t)) + 2M$ and we have the following estimate on the total variation of the solution:

$$(1.1.11) \qquad \begin{aligned} X(I(t)) &\leq \tfrac{2|I(t)|}{t} + 2M, \\ X(I(t)) &\equiv X_-(I(t)) + X_+(I(t)). \end{aligned}$$

Thus the nonlinearity $f''(u) \neq 0$ implies that the solution $u(x,t)$ is locally of *bounded variation* for any positive time $t > 0$. This is a very strong *compactness* property. Another simple corollary is that a periodic solution in x with period L decays over the period at a rate *independent* of the initial value:

$$(1.1.12) \qquad X(I) \leq \frac{2L}{t}.$$

This follows immediately from (1.1.10) by noting that the increasing and decreasing variations over a period are the same. There are interesting consequences of the strong convex nonlinearity; for instance, solutions with compact support have exactly two time invariants and tend to N-waves [12].

A weaker regularization property also holds for nonconvex flux functions. For this, though, we need a more general *Oleinik entropy condition* [25]:

$$(1.1.13) \qquad \frac{f(u) - f(u_-)}{u - u_-} \geq \frac{f(u_+) - f(u_-)}{u_+ - u_-} \qquad \text{for all } u \text{ between } u_- \text{ and } u_+.$$

This condition is related directly to viscosity, as we shall see in the next chapter. For now we simply mention that in general the solution of the Riemann problem is a wave form consisting of both shocks and rarefaction waves; see Figures 1.1.6 and 1.1.7.

1.2   Riemann Problem

We now consider the system of hyperbolic conservation laws $\mathbf{u} \in \mathbb{R}^n$, $n > 1$. The rarefaction waves and shock waves are vector valued, and the situation is more complicated than that for the scalar case considered above. A class of *simple waves* can be defined with properties similar to the rarefaction wave (1.1.8) for the scalar case

$$\mathbf{u}(x,t) = \phi(\xi(x,t)),$$

for a scalar function $\xi(x,t)$. Plug this into the conservation laws (1.1.1) and we have

$$\phi'(\xi(x,t))\xi_t(x,t) + f'(\phi(\xi(x,t)))\phi'(\xi(x,t))\xi_x(x,t) = 0.$$

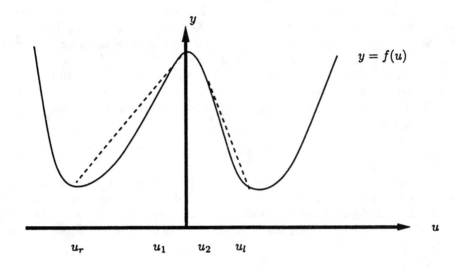

Figure 1.1.6. Riemann problem for nonconvex $f(u)$, phase space.

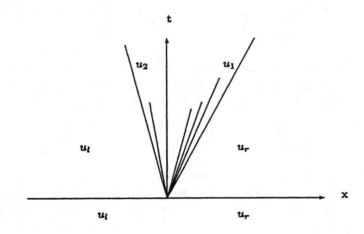

Figure 1.1.7. Riemann problem for nonconvex $f(u)$, physical space.

This implies that ϕ' is a right eigenvector r_i of f' and the speed of propagation $dx/dt = -\xi_t/\xi_x$ is an eigenvalue λ_i for some i, $1 \le i \le n$. We call this an *i-simple wave*. (In fact, this construction, in the infinitesimal sense, prompts the notion of hyperbolicity. Here we are actually going to construct not infinitesimal waves but nonlinear waves of finite strength.) The simple wave takes values along an integral curve of the vector field $r_i(\mathbf{u})$ in the state space. For a simple wave to exist in positive time the speed of propagation of the wave on the left has to be smaller than those on the right so as not to create the compression. As in the scalar case, for this to happen we need the eigenvalue $\lambda_i(\mathbf{u})$ to be strictly monotone along the integral curves of $r_i(\mathbf{u})$:

$$(1.2.1) \qquad \bigtriangledown \lambda_i(\mathbf{u}) \cdot r_i(\mathbf{u}) > 0.$$

Here we have chosen a preferred direction of the eigenvector. Condition (1.2.1) is reduced to the convexity condition $f''(u) \ne 0$ for the scalar equation. This condition for the general system is the *genuinely nonlinear* condition of Lax [11]. With this, we may divide a given integral curve $R_i(\mathbf{u}_0)$ of $r_i(\mathbf{u})$ through a given state $\mathbf{u}_0$ into

$$(1.2.2) \qquad R_i(\mathbf{u}_0) \equiv R_i^-(\mathbf{u}_0) \ U \ R_i^+(\mathbf{u}_0)$$

so that $\lambda_i(\mathbf{u}) < \lambda_i(\mathbf{u}_0)$ for the states $\mathbf{u} \in R_i^-(\mathbf{u}_0)$ and $\lambda_i(\mathbf{u}) \ge \lambda_i(\mathbf{u}_0)$ for $\mathbf{u} \in R_i^+(\mathbf{u}_0)$. Thus for a state $\mathbf{u}_2 \in R_i^+(\mathbf{u}_1)$ we can construct an *i-rarefaction wave* connecting $\mathbf{u}_1$ on the left and $\mathbf{u}_2$ on the right; cf. (1.1.8) and Figure 1.1.1 for the scalar law:

$$(1.2.3) \qquad \lambda_i((\mathbf{u}(x,t)) = \begin{cases} \lambda_i(\mathbf{u}_1), & x < \lambda_i(\mathbf{u}_1)t, \\ \frac{x}{t}, & \lambda_i(\mathbf{u}_1)t < x < \lambda_i(\mathbf{u}_2)t, \\ \lambda_i(\mathbf{u}_2), & x > \lambda_i(\mathbf{u}_2)t, \end{cases}$$
$$\mathbf{u}(x,t) \in R_i^+(\mathbf{u}_1).$$

We next investigate the discontinuity waves. The jump condition (1.1.6) is a vector relation and is not satisfied in general for two given states $\mathbf{u}_-$ and $\mathbf{u}_+$ even if we allow the speed s to vary. Our first task is then to investigate the *Hugoniot curves*

$$(1.2.4) \qquad H(\mathbf{u}_0) \equiv \{\mathbf{u} \ : \ \sigma(\mathbf{u} - \mathbf{u}_0) = f(\mathbf{u}) - f(\mathbf{u}_0)\}$$

for some scalar $\sigma = \sigma(\mathbf{u}_0, \mathbf{u})$. The jump condition (1.1.6) says that $\mathbf{u}_+ \in H(\mathbf{u}_-)$ and that $s = \sigma(\mathbf{u}_-, \mathbf{u}_+)$. When (1.1.1) is a linear system, $f'(\mathbf{u}) = A$ is an $n \times n$ constant matrix, σ is an eigenvalue, and $\mathbf{u} - \mathbf{u}_0$, $\mathbf{u} \in H(\mathbf{u}_0)$, is an eigenvector of A. In this case, the Hugoniot curves are actually lines in the eigenvector directions. The following basic theorem of [11] for general nonlinear systems is a consequence of the implicit function theorem.

THEOREM 1.2.1. *Suppose that system (1.1.1) is strictly hyperbolic. Then, in a small neighborhood of a state $\mathbf{u}_0$, the Hugoniot curves consist of n curves $H_i(\mathbf{u}_0)$, $i = 1, 2, \ldots, n$, with the following properties:*

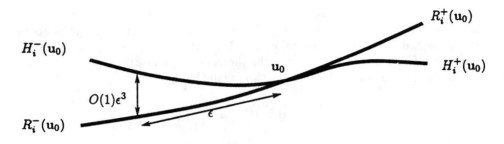

Figure 1.2.1. Wave curves.

(i) *The Hugoniot curve $H_i(\mathbf{u}_0)$ and the rarefaction curve $R_i(\mathbf{u}_0)$ have second-order contact at $\mathbf{u} = \mathbf{u}_0$; see Figure 1.2.1.*

(ii) *The shock speed $\sigma(\mathbf{u}_0, \mathbf{u})$, $\mathbf{u} \in H_i(\mathbf{u}_0)$, tends to the characteristic speed $\lambda_i(\mathbf{u}_0)$ as $\mathbf{u}$ tends to $\mathbf{u}_0$. Moreover,*

$$\sigma(\mathbf{u}_0, \mathbf{u}) = \frac{1}{2}(\lambda_i(\mathbf{u}) + \lambda_i(\mathbf{u}_0)) + O(1)|\mathbf{u} - \mathbf{u}_0|^2.$$

Proof. The existence of the Hugoniot curves is a consequence of the implicit function theorem. The jump condition (1.2.3) is rewritten as

$$\sigma(\mathbf{u} - \mathbf{u}_0) = \int_0^1 \frac{d}{d\tau} f(\tau\mathbf{u} + (1 - \tau)\mathbf{u}_0)d\tau = A(\mathbf{u}_0, \mathbf{u})(\mathbf{u} - \mathbf{u}_0),$$

$$A(\mathbf{u}_0, \mathbf{u}) \equiv \int_0^1 f'(\tau\mathbf{u} + (1 - \tau)\mathbf{u}_0)d\tau.$$

Thus $\mathbf{u} - \mathbf{u}_0$ is a right eigenvector $r^i(\mathbf{u})$ and σ is an eigenvalue $\lambda^i(\mathbf{u})$ of the matrix $A(\mathbf{u}_0, \mathbf{u})$. Let the left eigenvectors be $l^1(\mathbf{u})$, $l^2(\mathbf{u}), \ldots, l^n(\mathbf{u})$. The jump condition says that the vector $\mathbf{u} - \mathbf{u}_0$ is a right eigenvector and σ is an eigenvalue of $A(\mathbf{u}_0, \mathbf{u})$. Thus

$$l^j(\mathbf{u}) \cdot \sigma(\mathbf{u} - \mathbf{u}_0) - \lambda^j(\mathbf{u})l^j(\mathbf{u}) \cdot (\mathbf{u} - \mathbf{u}_0) = 0, \qquad j \neq i.$$

This is a system of $n - 1$ equations with trivial solution $\mathbf{u} = \mathbf{u}_0$. It can be solved to yield a curve $H_i(\mathbf{u}_0)$ through the state $\mathbf{u}_0$ by the implicit function theorem. We need only to check that the Jacobian J at $\mathbf{u} = \mathbf{u}_0$ has rank $n - 1$. Indeed if we take the gradient of the above $n - 1$ equations and then evaluate them at $\mathbf{u} = \mathbf{u}_0$ we get

$$J = \begin{pmatrix} l^1(\mathbf{u}_0) \\ \cdot \\ \cdot \\ \cdot \\ l^{i-1}(\mathbf{u}_0) \\ l^{i+1}(\mathbf{u}_0) \\ \cdot \\ \cdot \\ l^n(\mathbf{u}_0) \end{pmatrix} \cdot$$

This has rank $n - 1$ because these vectors $l^k(\mathbf{u}_0) = l_k(\mathbf{u}_0)$, $k \neq i$, are linearly independent due to the assumed strict hyperbolicity of system (1.1.1). The above is a simplified presentation of Conlon [25]. From $A(\mathbf{u}_0, \mathbf{u}_0) = f'(\mathbf{u}_0)$, we have the first part of each of the assertions in the theorem. We now differentiate the jump condition with respect to the arc length μ along $H_i(\mathbf{u}_0)$:

$$\frac{d\sigma}{d\mu}(\mathbf{u} - \mathbf{u}_0) = (f'(\mathbf{u}) - \sigma)\frac{d\mathbf{u}}{d\mu}.$$

Evaluating this at $\mathbf{u} = \mathbf{u}_0$, we get $\sigma(\mathbf{u}_0, \mathbf{u}_0) = \lambda_i(\mathbf{u}_0)$ and $d\mathbf{u}/d\mu = r_i(\mathbf{u}_0)$ as shown above. Differentiating one more time and then evaluating again at $\mathbf{u} = \mathbf{u}_0$:

$$2\frac{d\sigma}{d\mu}r_i(\mathbf{u}_0) = f''(\mathbf{u}_0)r_i(\mathbf{u}_0)r_i(\mathbf{u}_0) + (f'(\mathbf{u}_0) - \lambda_i(\mathbf{u}_0))\frac{d^2\mathbf{u}}{d\mu^2}.$$

Next we differentiate the relation $f'(\mathbf{u})r_i(\mathbf{u}) = \lambda_i(\mathbf{u})r_i(\mathbf{u})$ with respect to the arc length ν along the rarefaction curve $R_i(\mathbf{u}_0)$ and evaluate it also at $\mathbf{u} = \mathbf{u}_0$:

$$\frac{d\lambda_i}{d\nu}r_i(\mathbf{u}_0) = f''(\mathbf{u}_0)r_i(\mathbf{u}_0)r_i(\mathbf{u}_0) - (\lambda_i(\mathbf{u}_0) - f'(\mathbf{u}_0))\frac{dr_i}{d\nu}.$$

Since we already know that $d/d\mu = d/d\nu$ at $\mathbf{u} = \mathbf{u}_0$, we have from subtracting the above identities that

$$\frac{d(2\sigma - \lambda_i)}{d\mu}r_i(\mathbf{u}_0) = (f'(\mathbf{u}_0) - \lambda_i(\mathbf{u}_0))\left(\frac{d^2\mathbf{u}}{d\mu^2} - \frac{dr_i}{d\mu}\right).$$

Since the operation $f' - \lambda_i$ annihilates the r_i component of any vector, both sides of the above identity are zero. In particular,

$$\frac{d\sigma}{d\mu} = \frac{1}{2}\frac{d\lambda_i}{d\mu}$$

at $\mathbf{u} = \mathbf{u}_0$. This proves (ii) of the theorem. Let $\mathbf{u}'$ denote the differentiation at $\mathbf{u} = \mathbf{u}_0$ along $H_i(\mathbf{u}_0)$. We have from the above two identities that, at $\mathbf{u} = \mathbf{u}_0$,

$$(f' - \lambda_i)(\mathbf{u}'' - r_i') = 0.$$

Since the differentiation is with respect to the arc length μ, the vector $\mathbf{u}' = r_i$ is a unit vector. Thus $\mathbf{u}''$, r_i' are orthogonal to r_i. Writing them as a linear combination of the eigenvectors,

$$\mathbf{u}'' - r_i' = \sum_j c_j r_j,$$

we have

$$0 = (f' - \lambda_i)(\mathbf{u}'' - r_i') = (\lambda_i - f')\sum_j c_j r_j = \sum_j c_j(\lambda_j - \lambda_i)r_j.$$

This implies that $c_j = 0$, $j \neq i$, and so $\mathbf{u}'' - r_i{}'$ is parallel to r_i. But we already know that it is orthogonal to r_i. Thus we conclude that $\mathbf{u}'' = r_i{}'$, which yields the second statement in (i). This completes the proof of the theorem.   □

From part (ii) of Theorem 1.2.1, we may partition the Hugoniot curve in the same way as the rarefaction curves, (1.2.2):

$$(1.2.5) \qquad\qquad H_i(\mathbf{u}_0) = H_i^-(\mathbf{u}_0) \ U \ H_i^+(\mathbf{u}_0),$$

so that $\lambda_i(\mathbf{u}) < \lambda_i(\mathbf{u}_0)$ for the states $\mathbf{u} \in H_i^-(\mathbf{u}_0)$ and $\lambda_i(\mathbf{u}) \geq \lambda_i(\mathbf{u}_0)$ for $\mathbf{u} \in H_i^+(\mathbf{u}_0)$. We say that an i-characteristic field is *linearly degenerate* in the sense of Lax if

$$(1.2.6) \qquad\qquad \bigtriangledown \lambda_i(\mathbf{u}) \cdot r_i(\mathbf{u}) \equiv 0.$$

COROLLARY 1.2.1. *Suppose that system (1.1.1) is strictly hyperbolic and that each characteristic field is either genuinely nonlinear, (1.2.1), or linearly degenerate, (1.2.6). Then in a small neighborhood of a given state $\mathbf{u}_0$, we have the following conditions:*

(i) *For a genuinely nonlinear field, $(\mathbf{u}_0, \mathbf{u})$, $\mathbf{u} \in R_i^+(\mathbf{u}_0)$, is a rarefaction wave, and $(\mathbf{u}_-, \mathbf{u}_+) = (\mathbf{u}_0, \mathbf{u})$, $\mathbf{u} \in H_i^-(\mathbf{u}_0)$, is a shock satisfying the Lax entropy condition*

$$(1.2.7) \qquad\qquad \lambda_i(\mathbf{u}_-) > s > \lambda_i(\mathbf{u}_+).$$

(ii) *For a linearly degenerate field, $H_i(\mathbf{u}_0) = R_i(\mathbf{u}_0)$ and $(\mathbf{u}_0, \mathbf{u})$, $\mathbf{u} \in R_i(\mathbf{u}_0)$, form a contact discontinuity with speed s:*

$$(1.2.8) \qquad\qquad s = \lambda_i(\mathbf{u}) = \lambda_i(\mathbf{u}_0).$$

Proof. Part (i) follows from Theorem 1.2.1. To prove (ii) we first notice that along $R_i(\mathbf{u}_0)$, $(d/d\nu)\lambda_i(\mathbf{u}) = 0$ by (1.2.6). From this we notice that the jump condition is satisfied along $R_i(\mathbf{u}_0)$ with $\sigma(\mathbf{u}_0, \mathbf{u}) = \lambda_i(\mathbf{u}) = \lambda_i(\mathbf{u}_0)$:

$$\frac{d}{d\nu}[\lambda_i(\mathbf{u})(\mathbf{u} - \mathbf{u}_0) - (f(\mathbf{u}) - f(\mathbf{u}_0))] = \lambda_i(\mathbf{u})r_i(\mathbf{u}) - f'(\mathbf{u})r_i(\mathbf{u}) = 0. \qquad □$$

From Corollary 1.2.1, we construct the *wave curves* $W_i(\mathbf{u}_0)$ as follows:

$$(1.2.9)$$
$$W_i(\mathbf{u}_0) \equiv \begin{cases} R_i^+(\mathbf{u}_0) \ U \ H_i^-(\mathbf{u}_0), & \text{ith characteristic genuinely nonlinear;} \\ R_i(\mathbf{u}_0) = H_i(\mathbf{u}_0), & \text{ith characteristic linearly degenerate.} \end{cases}$$

Thus $(\mathbf{u}_0, \mathbf{u})$ forms an elementary i-wave described above when $\mathbf{u} \in W_i(\mathbf{u}_0)$. These waves are the building blocks for the solution of the Riemann problem for (1.1.1). An important consequence of part (i) of Theorem 1.2.1 is that the wave curves are C^2 curves.

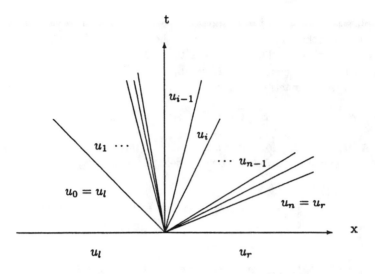

Figure 1.2.2. Solution to the Riemann problem.

THEOREM 1.2.2 (see [2]). *Suppose that each characteristic field is either genuinely nonlinear or linearly degenerate. Then the Riemann problem for (1.1.1) has a unique solution in the class of elementary waves provided that the states are in a small neighborhood of some state* $\mathbf{u}_0$.

Proof. We need to show that for the given states $\mathbf{u}_l$, $\mathbf{u}_r$ there exist unique states $\mathbf{u}_0 = \mathbf{u}_l$, $\mathbf{u}_n = \mathbf{u}_r$, $\mathbf{u}_i \in W_i(\mathbf{u}_{i-1})$, $i = 1, 2, \ldots, n$; see Figure 1.2.2. Parametrize each wave curve W_i by arc length μ_i. We construct a function F that maps a small neighborhood of zero in $\mathbb{R}^n$ into a small neighborhood of $\mathbf{u}_l$ in the state space

$$F(y_1, y_2, \ldots, y_n) = \mathbf{u},$$

with the property that there exist states $\mathbf{u}_i$, $i = 0, 1, \ldots, n$, $\mathbf{u}_0 = \mathbf{u}_l$, $\mathbf{u} = \mathbf{u}_n$, $\mathbf{u}_i \in W_i(\mathbf{u}_{i-1})$, $\mu_i(\mathbf{u}_i) - \mu_i(\mathbf{u}_{i-1}) = y_i$, $i = 1, \ldots, n$. It suffices to show that the mapping $F(x)$ is invertible in a small neighborhood of zero. Notice from part (i) of Theorem 1.2.1 that the wave curves are C^2. Thus we may apply the inverse function theorem for which we need to check its Jacobian at $x = 0$. We know from part (i) of Theorem 1.2.1 and Corollary 1.2.1 that the wave curve $W_i(\mathbf{u}_0)$ has tangent $r_i(\mathbf{u}_0)$ at $\mathbf{u} = \mathbf{u}_0$. Thus it is clear from the construction above that $\frac{\partial}{\partial y_i} F(y)|_{y=0} = r_i(\mathbf{u}_0)$. Since $\{r_1(\mathbf{u}_0), \ldots, r_n(\mathbf{u}_0)\}$ are linearly independent as a consequence of strict hyperbolicity, the Jacobian of F has rank n and the theorem is proved.   □

The above has been generalized to systems that are strictly hyperbolic but not necessarily genuinely nonlinear or linearly degenerate. For the general situation we need an entropy condition [16] generalizing that of Lax.

DEFINITION 1.2.1. *A shock* $(\mathbf{u}_-, \mathbf{u}_+)$, $\mathbf{u}_+ \in H_i(\mathbf{u}_-)$, *with speed* $s = \sigma(\mathbf{u}_-, \mathbf{u}_+)$

is admissible if it satisfies the following generalized entropy condition:

(1.2.10)
$$s = \sigma(\mathbf{u}_-, \mathbf{u}_+) \leq \sigma(\mathbf{u}_-, \mathbf{u}) \qquad \text{for any } \mathbf{u} \in H_i(\mathbf{u}_-) \text{ between } \mathbf{u}_- \text{ and } \mathbf{u}_+.$$

It can be shown that, as in Theorem 1.2.2, the Riemann problem for a general strictly hyperbolic system can be solved uniquely in the class of elementary waves under the above entropy condition. The construction of wave curves involves the combination of Hugoniot and rarefaction curves; cf. Figure 1.1.7 for a scalar equation.

The Riemann problem is important for several reasons. Because its initial data as well as the hyperbolic conservation laws are invariant under the dilation $x \rightarrow cx$, $t \rightarrow ct$, assuming the uniqueness of the initial value problem, the solution is a function of x/t. Thus it is much easier to solve than the general initial value problem. On the other hand, its solution requires the identification of the entropy condition and yields the elementary waves, which can form the building blocks for general solutions.

Another important aspect of the solution of the Riemann problem is that it forms a *noninteracting* wave pattern in that the elementary waves diverge and do not interact in the positive time direction. A general solution of (1.1.1) contains complex nonlinear wave interactions. These interactions would combine and cancel waves, as simple examples for scalar equations in the last chapter indicate; see Figures 1.1.3 and 1.1.4. For systems, the interactions also give rise to new waves. Nevertheless, the solution would, time asymptotically, tend to a wave pattern that is noninteracting. Thus the Riemann solution would represent the *scattering data* for a general solution of (1.1.1). This fact will be shown in section 1.4 for a general class of initial data.

In fact, the Riemann solution also represents the local behavior of a general solution. This is so locally in time, because, except for countable points of point interaction, there are few interactions in a small neighborhood of a given point, and waves also scatter out in the forward time direction. This is the regularity result to be studied also in section 1.4. The following proposition on the almost noninteracting wave pattern of a given characteristic family is useful for these studies.

PROPOSITION 1.2.1. *Suppose that the ith characteristic field is genuinely nonlinear. Consider a wave pattern $\mathbf{W}$ of weak i-waves connecting the states $\mathbf{u}_l$ to the left and $\mathbf{u}_r$ to the right, of finite total strength and almost divergent in the forward time direction in the sense that $\lambda_i(\mathbf{u}(x_1)) - \lambda_i(\mathbf{u}(x_2)) < \delta$ for $x_1 < x_2$. Then, for δ sufficiently small, $\mathbf{W}$ takes values in a δ neighborhood of the rarefaction curve $R_i(\mathbf{u}_l)$ and $\mathbf{u}(x_2) \in R_i^+(\mathbf{u}(x_1)) + O(1)\delta$ for $x_2 > x_1$.*

Proof. Since an i-shock of strength ε takes values along a rarefaction curve up to the error $O(1)\varepsilon^3$, Theorem 1.2.1, it suffices to show that the total error is small:

$$\sum \{|\alpha|^3 : \alpha \text{ a shock in } \mathbf{W}\} < \delta.$$

From the divergence hypothesis, we know that each i-shock in $\mathbf{W}$ is of strength no greater than $O(1)\delta$. Let the total strength of $\mathbf{W}$ be M; then

$$\sum\{|\alpha|^3 : \alpha \text{ a shock in } \mathbf{W}\} = (O(1)\delta)^2 M = O(1)\delta^2 < \delta.$$

This proves the first statement; the second follows from the first and the divergence hypothesis.   □

1.3   Wave Interactions

We have seen for the convex scalar equation that waves either combine or cancel. For a system the interaction of waves pertaining to the same characteristic family in general produce waves of other characteristic families. The interaction of a shock and a rarefaction wave of the same family may produce infinitely many shocks in finite time. In this section we will study only the relation of the waves before the interaction and the scattering data for the completed interaction. The scattering data, as we have seen in the last section, are the solution of the Riemann problem. We will consider the interaction of two sets of solutions of the Riemann problem.

THEOREM 1.3.1. *Suppose that (1.1.1) is strictly hyperbolic and that each characteristic family is either genuinely nonlinear or linearly degenerate. Let $\mathbf{u}_l$, $\mathbf{u}_m$, and $\mathbf{u}_r$ be three nearby states and the solutions of the Riemann problems $(\mathbf{u}_l, \mathbf{u}_m)$ and $(\mathbf{u}_m, \mathbf{u}_r)$ be $(\mathbf{v}_{i-1}, \mathbf{v}_i)$ and $(\mathbf{w}_{i-1}, \mathbf{w}_i)$, $i = 1, 2, \ldots, n$, respectively; see Figure 1.3.1. Then the solution $(\mathbf{u}_{i-1}, \mathbf{u}_i)$, $i = 1, 2, \ldots, n$, of the Riemann problem $(\mathbf{u}_l, \mathbf{u}_r)$ is the linear superposition of the above two solutions modulo the nonlinear effect of the order $Q(\mathbf{u}_l, \mathbf{u}_m, \mathbf{u}_r) = \mathbf{Q}(\mathbf{W})$, the degree of interaction for the wave pattern $\mathbf{W}$ consisting of the solutions of the Riemann problems $(\mathbf{u}_l, \mathbf{u}_m)$ on the left and $(\mathbf{u}_m, \mathbf{u}_r)$ on the right. In other words,*

$$\bar{\mathbf{u}}_i = \bar{\mathbf{v}}_i + \bar{\mathbf{w}}_i + O(1)Q(\mathbf{u}_l, \mathbf{u}_m, \mathbf{u}_r),$$

$$\bar{\mathbf{u}}_i \equiv \mathbf{u}_i - \mathbf{u}_{i-1}, \quad \bar{\mathbf{v}}_i \equiv \mathbf{v}_i - \mathbf{v}_{i-1}, \quad \bar{\mathbf{w}}_i \equiv \mathbf{w}_i - \mathbf{w}_{i-1},$$

$$Q(\mathbf{u}_l, \mathbf{u}_m, \mathbf{u}_r) = Q_s(\mathbf{u}_l, \mathbf{u}_m, \mathbf{u}_r) + Q_d(\mathbf{u}_l, \mathbf{u}, m, \mathbf{u}_r),$$

(1.3.1) $$Q_s(\mathbf{u}_l, \mathbf{u}_m, \mathbf{u}_r) \equiv \sum_{i=} \{\alpha_i \beta_i (\alpha_i + \beta_i) : \text{ one or both of } (\mathbf{v}_{i-1}, \mathbf{v}_i)$$

$$\text{and } (\mathbf{w}_{i-1}, \mathbf{w}_i) \text{ shocks}\},$$

$$Q_d(\mathbf{u}_l, \mathbf{u}_m, \mathbf{u}_r) \equiv \sum_{i>j} \alpha_i \beta_j,$$

$$\alpha_i \equiv |\mathbf{v}_i - \mathbf{v}_{i-1}|, \quad \beta_i \equiv |\mathbf{w}_i - \mathbf{w}_{i-1}|, \quad \gamma_i \equiv |\mathbf{u}_i - \mathbf{u}_{i-1}|$$

for bounded functions $O(1)$ whose bounds depend only on the flux $f(\mathbf{u})$.

Proof. The proof depends on the fact that the wave curves $W_i(\mathbf{u}_0)$, (1.2.9), are C^2 and is done in steps. The trivial case where $Q(\mathbf{u}_l, \mathbf{u}_m, \mathbf{u}_r) = 0$ occurs if and only if, according to the above definition, there exists k, $1 \leq k \leq n$, such that $\alpha_k = \alpha_{k+1} = \cdots = \alpha_n = 0$, $\beta_1 = \beta_2 = \cdots = \beta_{k-1} = 0$, and, moreover, both (v_{k-1}, v_k) and (w_k, w_{k-1}) are rarefaction waves. In this case, the solution of the Riemann problem $(\mathbf{u}_l, \mathbf{u}_r)$ is simply the combination of those of $(\mathbf{u}_l, \mathbf{u}_m)$ and $(\mathbf{u}_m, \mathbf{u}_r)$ and there is no nonlinear interaction; see Figure 1.3.2.

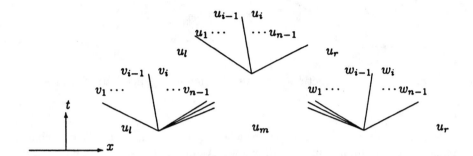

Figure 1.3.1. Wave interactions.

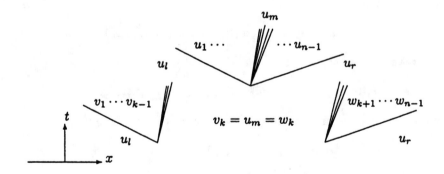

Figure 1.3.2. Noninteracting wave pattern.

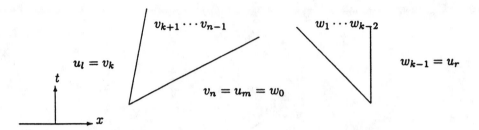

Figure 1.3.3. Interaction of waves of different characteristic families.

We now consider the nontrivial cases where $Q(\mathbf{u}_l, \mathbf{u}_m, \mathbf{u}_r) \neq 0$. We first consider the data with $Q_s(\mathbf{u}_l, \mathbf{u}_m, \mathbf{u}_r) = 0$. Thus we assume that there exists k, $1 \leq k \leq n$, such that there is no wave faster (or slower) than k-waves to the right (or left); see Figure 1.3.3:

$\alpha_1 = \cdots = \alpha_k = 0$, $\beta_{k+1} = \cdots = \beta_n = 0$. We have

$$Q_d(\mathbf{u}_l, \mathbf{u}_m, \mathbf{u}_r) = |\alpha||\beta|, \qquad |\alpha| \equiv \sum_i \alpha_i, \qquad |\beta| \equiv \sum_i \beta_i.$$

If $|\alpha|$ (or $|\beta|$) is zero, then $\mathbf{u}_i = \mathbf{w}_1$ (or $\mathbf{u}_i = \mathbf{v}_i$) and the solution of the Riemann problem $(\mathbf{u}_l, \mathbf{u}_r)$ is the same as that of $(\mathbf{u}_l, \mathbf{u}_m)$ (or $(\mathbf{u}_m, \mathbf{u}_r)$). In this case the theorem holds trivially with $Q(\mathbf{u}_l, \mathbf{u}_m, \mathbf{u}_r) = 0$. Since the solution of a Riemann problem is a C^2 function of its data, the functions

$$\theta_i(\alpha, \beta) \equiv \gamma_i - \alpha_i - \beta_i, \qquad i = 1, 2, \ldots, n,$$

are C^2 functions. Here

$$\alpha \equiv (\alpha_{k+1}, \ldots, \alpha_n), \qquad \beta \equiv (\beta_1, \ldots, \beta_k).$$

The above analysis shows that

$$\theta_i(0, \ldots, 0; \beta_1, \ldots, \beta_k) = \theta_i(\alpha_{k+1}, \ldots, \alpha_n; 0, \ldots, 0) = 0.$$

By the intermediate function theorem we deduce that

$$\theta_i(\alpha, \beta) = \theta_i(\alpha, \beta) - \theta_i(0, \beta) = D_\alpha \theta_i(\bar{\alpha}, \beta) \cdot \alpha$$

for some state $\bar{\alpha}$ between $\mathbf{0}$ and α. Similarly,

$$\begin{aligned} D_\alpha \theta_i(\bar{\alpha}, \beta) \cdot \alpha &= (D_\alpha \theta_i(\bar{\alpha}, \beta) - D_\alpha \theta_i(\bar{\alpha}, 0)) \cdot \alpha \\ &= D_\beta D_\alpha \theta_i(\bar{\alpha}, \bar{\beta}) \cdot \alpha \cdot \beta = O(1)|\alpha||\beta|. \end{aligned}$$

Thus we conclude that $\theta_i = O(1)|\alpha||\beta| = O(1)Q_d(\mathbf{u}_l, \mathbf{u}_m, \mathbf{u}_r)$ and the theorem holds in this case.

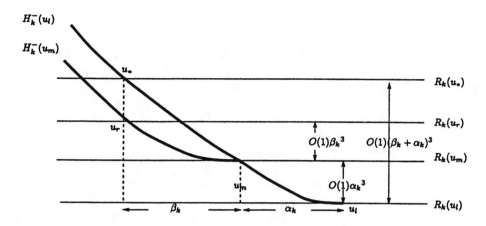

Figure 1.3.4. Interaction of shocks, phase plane.

Next we consider the second type of interaction, the interaction of the two k-waves α_k and β_k for some k, $1 \le k \le n$, and assume that there is no other wave: $\alpha_i = \beta_i = 0$ for $i \ne k$. If the two k-waves are both rarefaction waves, then this case is a special case of the noninteracting wave pattern considered at the beginning. We have $Q_s(\mathbf{u}_l, \mathbf{u}_m, \mathbf{u}_r) = 0$ and the result of the interaction is simply the linear superposition of the rarefaction waves. The theorem holds trivially in this case. When one of the waves is a shock wave, $Q_s(\mathbf{u}_l, \mathbf{u}_m, \mathbf{u}_r) = \alpha_k \beta_k (\alpha_k + \beta_k)$. Take the case of two shocks that do not simply combine because the Hugoniot curves depend on the initial state: $\mathbf{u}_m \in H_k(\mathbf{u}_l)$, $\mathbf{u}_r \in H_k(\mathbf{u}_m)$, but in general $\mathbf{u}_r$ is not on $H_k(\mathbf{u}_l)$; see Figure 1.3.4.

By part (i) of Theorem 1.2.1 there exists $\mathbf{u}_*$ on $H_k(\mathbf{u}_l)$, which is of distance $O(1)((\alpha_k + \beta_k)^3 - \alpha_k{}^3 - \beta_k{}^3) = O(1)Q_s(\mathbf{u}_l, \mathbf{u}_m, \mathbf{u}_r)$ from the state $\mathbf{u}_r$; see Figure 1.3.5.

We now view the Riemann problem $(\mathbf{u}_l, \mathbf{u}_r)$ as the result of the interaction of a k-shock $(\mathbf{u}_l, \mathbf{u}_*)$ and the solution of the Riemann problem $(\mathbf{u}_*, \mathbf{u}_r)$; see Figure 1.3.5. Since the jump $|\mathbf{u}_* - \mathbf{u}_r|$ is of the order $O(1)Q_s$, the solution of the Riemann problem $(\mathbf{u}_l, \mathbf{u}_r)$ is an $O(1)Q_s$ perturbation of the shock $(\mathbf{u}_l, \mathbf{u}_*)$ and the theorem follows. The interaction of two other types of k-waves is similar.

Finally the general interactions are reduced to a series of interactions of the above two types plus interaction with waves of strength of the order $Q(\mathbf{u}_l, \mathbf{u}_m, \mathbf{u}_r)$. For simplicity, we will consider two conservation laws, $n = 2$.

Consider first the interaction of the 2-wave $(\mathbf{v}_1, \mathbf{v}_2)$ to the left with the 1-waves $(\mathbf{w}_0, \mathbf{w}_1)$ to the right; see the first and second rows of Figure 1.3.6.

This interaction is of the first type we studied first in this proof and the error to the linear superposition is $O(1)\alpha_2\beta_1 = O(1)Q_d(\mathbf{u}_l, \mathbf{u}_m, \mathbf{u}_r)$. The interaction yields the solution of the Riemann problem $(\mathbf{v}_2, \mathbf{w}_1$ consists of 1-waves θ_1 and 2-wave θ_2 satisfying

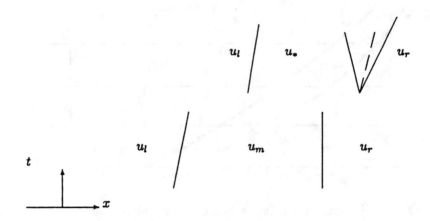

Figure 1.3.5. Interaction of shocks, physical space.

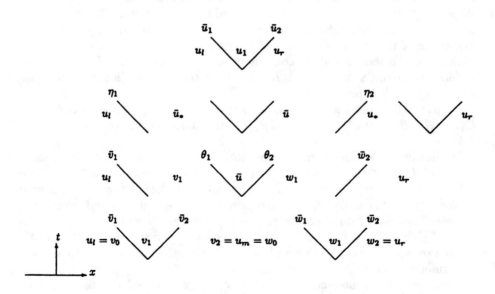

Figure 1.3.6. General interaction of waves, $u \in R^2$.

$$(1.3.2) \qquad \theta_i = \begin{cases} (\mathbf{w}_1, \mathbf{w}_i) + O(1)Q_d(\mathbf{u}_l, \mathbf{u}_m, \mathbf{u}_r), & i = 1, \\ (\mathbf{v}_2, \mathbf{v}_n) + O(1)Q_d(\mathbf{u}_l, \mathbf{u}_m, \mathbf{u}_r), & i = 2. \end{cases}$$

These waves are separated by the state $\bar{u}$; see Figure 1.3.6. Here and in what follows, a wave denotes either its end states or the jump of its states, whichever is convenient for us. This causes no confusion. We next let the two 1-waves $\bar{\mathbf{v}}_1$ and θ_1 interact. This is the second type of interaction considered above. It results in the linear superposition $\eta_1 = (\mathbf{u}_l, \bar{\mathbf{u}}_*)$ of the 1-waves plus an error

$$|\bar{\mathbf{u}}_* - \bar{\mathbf{u}}| = O(1)Q_s(\mathbf{u}_l, \mathbf{u}_1, \bar{\mathbf{u}});$$

see the second and third rows of Figure 1.3.6. Note that from (1.3.2) the interaction potential $Q_s(\mathbf{u}_l, \mathbf{u}_1, \bar{\mathbf{u}})$ between the waves $\bar{\mathbf{v}}_1$ and θ_1 is the same as that between $\bar{\mathbf{v}}_1$ and $\bar{\mathbf{w}}_1$ plus an error of the order $Q_d(\mathbf{u}_l, \mathbf{u}_m, \mathbf{u}_r)$. Thus we have from the above estimate that

$$|\bar{\mathbf{u}}_* - \bar{\mathbf{u}}| = O(1)Q(\mathbf{u}_l, \mathbf{u}_m, \mathbf{u}_r).$$

Similarly, we let the 2-waves θ_2 and $\mathbf{w}_2$ interact and obtain the error

$$|\mathbf{u}_* - \mathbf{u}_r| = O(1)Q_s(\bar{\mathbf{u}}, \mathbf{w}_1, \mathbf{u}_r) = O(1)Q(\mathbf{u}_l, \mathbf{u}_m, \mathbf{u}_r)$$

to the linear superposition $\eta_2 = (\bar{\mathbf{u}}, \mathbf{u}_*)$; see the third row of Figure 1.3.6. The final step is to view the solution of the Riemann problem (u_l, u_r) as the linear superposition of the noninteracting waves η_1 and η_2 plus a perturbation of the order of $|\bar{\mathbf{u}}_* - \bar{\mathbf{u}}| + |\mathbf{u}_* - \mathbf{u}_r| = O(1)Q(u_l, u_m, u_r)$. This completes the proof of the theorem.   □

1.4   Random Choice Method

Nonlinear interaction of waves can be controlled globally in many cases and solutions of general initial value problems can be constructed using the elementary waves studied in the last section as building blocks. This has been done for the general system

$$(1.4.1) \qquad\qquad \mathbf{u}_t + f(\mathbf{u})_x = 0$$

when the initial data have small total variation TV:

$$(1.4.2) \qquad \mathbf{u}(x, 0) = \mathbf{u}_0(x), \qquad TV \equiv \text{variation}_{-\infty < x < \infty} \, \mathbf{u}_0(x).$$

We now present the important work of Glimm. The Glimm scheme is a finite difference scheme involving a random sequence a_i, $i = 0, 1, \ldots$, $0 < a_i < 1$. Let $r = \Delta x$, $s = \Delta t$ be the mesh sizes satisfying the Courant–Friedrich–Lewy (CFL) condition

$$(1.4.3) \qquad\qquad \frac{r}{s} > 2|\lambda_i(\mathbf{u})|, \qquad 1 \le i \le n,$$

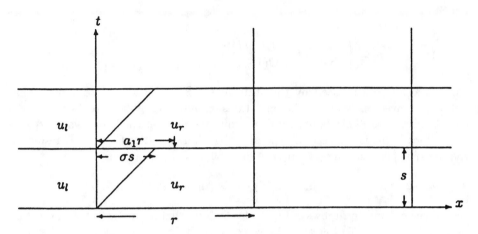

Figure 1.4.1. Random choice, $a_1 r > \sigma s$.

for all states **u** under consideration. The approximate solutions $\mathbf{u}(x,t) = \mathbf{u}_r(x,t)$ depend on the random sequence $\{a_k\}$ and are defined inductively in time as follows:

$$(1.4.4) \quad u(x,0) = \mathbf{u}_0((h + a_0)r), \qquad hr < x < (h+1)r,$$

$$(1.4.5) \quad \mathbf{u}(x,ks) = \mathbf{u}((h + a_k)r - 0, ks - 0), \qquad hr < x < (h+1)r,$$

$$k = 1, 2, \ldots .$$

Thus the approximate solution is a step function for each layer $t = ks$, $k = 0, 1, 2, \ldots$. Between the layers it consists of elementary waves obtained by solving the Riemann problems at each grid point $x = hr$, $h = 0, \pm 1, \ldots$. Due to the CFL condition these elementary waves do not interact within the layer. Thus the approximation solution is an exact solution except at the interfaces $t = 0, s, 2s, \ldots$. The CFL condition depends on the solution and needs to be checked at each time step. The numerical error depends on the random sequence. Take the example of the propagation of a single shock with positive speed σ:

$$\mathbf{u}_0(x) = \begin{cases} \mathbf{u}_-, & x < 0, \\ \mathbf{u}_+, & x > 0. \end{cases}$$

The shock is located at $x = \sigma s$ at $t = s - 0$. At $t = s$ it is located at

$$x = \begin{cases} 0 & \text{if } a_1 r > \sigma s, \\ r & \text{if } a_1 r \le \sigma s; \end{cases}$$

see Figures 1.4.1 and 1.4.2.

Given a fixed time $t = T = Ks$ the location of the shock in the approximate solution is

$$x = A(K, I)r, \qquad I \equiv \left(0, \sigma \frac{s}{r}\right).$$

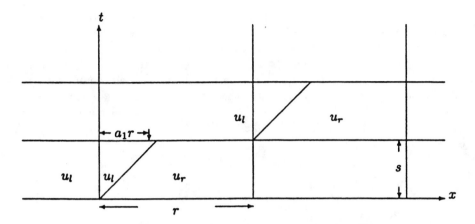

Figure 1.4.2. Random choice, $a_1 r < \sigma s$.

Here, for a given subinterval I of $(0,1)$ and positive integer N, $A(N,I)$ denotes the number of k, $1 \le k \le N$, such that $a_k \in I$. When the meshes r, s, r/s fixed, are refined, we have $K \to \infty$ and the shock location becomes exact at $x = \sigma T$ if

$$A(K,I)r \;\to\; \sigma T, \qquad \text{or } \tfrac{A(K,I)}{K|I|} \;\to\; 1 \qquad \text{as } K \to \infty.$$

Here $|I| = \sigma s/r$ is the length of the interval I. In other words, the shock location is exact in the limit if the sequence a_k is equidistributed. See Definition 1.4.1.

DEFINITION 1.4.1. *A sequence* $\{a_k\}_{k=1}^{\infty}$ *in* $(0,1)$ *is equidistributed if*

$$B(N,I) \equiv |\tfrac{A(N,I)}{N} - |I|| \to 0 \qquad \text{as } N \to \infty$$

for any subinterval I of $(0,1)$. Here $A(N,I)$ denotes the number of k, $1 \le k \le N$, such that $a_k \in I$, and $|I|$ is the length of I.

Almost all sequences are equidistributed. We have the following major result.

THEOREM 1.4.1 (see [5]). *Suppose that the initial data $\mathbf{u}_0(x)$ are of small total variation TV. Then the approximate solutions $\mathbf{u}(x,t)$ are of small total variation $O(1)TV$ in x for all time t. Moreover, for almost all choices of the sequence $\{a_k\}_{k=1}^{\infty}$, the approximate solutions tend to an exact solution for a sequence of the mesh sizes r, s tending to zero with r/s fixed and r, s satisfying the CFL condition. The exact solution $\mathbf{u}(x,t)$ is of bounded variation in x for any time $t \ge 0$:*

$$\text{variation}_{-\infty < x < \infty} \; \mathbf{u}(x,t) = O(1)TV$$

and is continuous in the $L_1(x)$-norm:

$$\int_{-\infty}^{\infty} |\mathbf{u}(x,t_1) - \mathbf{u}(x,t_2)|dx = O(1)|t_1 - t_2|, \qquad t_1, t_2 \ge 0.$$

Proof. This long and important proof is done in the following three steps.

Boundedness. This step is to prove the uniform boundedness of the total variation of the approximate solutions. The main idea is that of the *Glimm functional* $F(J)$ defined on space-like curves J. It consists of a linear part $L(J)$, measuring the total variation, and a quadratic part $Q(J)$, measuring the potential wave interaction. The curve J incorporates the scheme and consists of line segments connecting points $((h \pm a_k)r, ks)$ and $(hr, (k \pm 1/2)s)$. The elementary waves issued from the grid points (hr, ks) will cross the line segments. The functionals are defined as follows:

$$
\begin{aligned}
&L(J) \equiv \sum \{|\alpha| : \; \alpha \text{ any wave crossing } J\}, \\
&Q_d(J) \equiv \sum \{|\alpha||\beta| : \; \alpha \text{ and } \beta \\
&\quad \text{interacting waves of distinct characteristic families crossing } J\}, \\
&Q_s(J) \equiv \sum_{i=1}^{n} Q_s^i, \\
&Q_s^i \equiv \sum \{|\alpha||\beta|(-\min\{\Theta(\alpha, \beta), 0\}) : \; \alpha \text{ and } \beta \\
&\quad \text{interacting } i\text{-waves crossing } J\}, \\
&Q(J) \equiv Q_d(J) + Q_s(J), \\
&F(J) \equiv L(J) + MQ(J).
\end{aligned}
$$

(1.4.6)

Here M is a sufficiently large constant to be chosen later. In the above definition, the notion of *interacting* is analogous to those in Q_s and Q_d, (1.3.1): An i-wave to the left of a j-wave is interacting if either $i > j$ or $i = j$ and at least one of them is a shock wave. The term $\Theta(\alpha, \beta)$ for two i-waves represents the effective angle between the waves and is set as follows (for definiteness, we assume that α lies to the left of β):

(1.4.7) $$\Theta(\alpha, \beta) \equiv \theta_\alpha^+ + \theta_\beta^- + \sum \theta_\delta.$$

Here θ_α^+ represents the speed of α minus the value of λ_i at its right state. It is negative if α is a shock and is set to zero if it is an i-rarefaction wave. Similarly the term θ_β^- denotes the difference between the speed of β and the i-characteristic value of its left end state. θ_δ is the value of λ_i at the right state of the wave δ minus that at the left state. It is positive if δ is a rarefaction wave and is negative if it is a shock. The sum $\sum \theta_\delta$ is over the i-waves δ between α and β; see Figure 1.4.3.

Subject to wave interactions of distinct families, $-\Theta(\alpha, \beta)$ represents the effective angle between the i-waves α and β when waves of other characteristic families between them propagate away. When $\Theta(\alpha, \beta)$ is positive, the two waves are not likely to meet and should not be included in the functional Q_s. When $\Theta(\alpha, \beta)$ is negative, the two waves will eventually meet and interact. Thus $Q_s(J)$ reflects accurately the potential interactions of waves of the same characteristic family.

The main estimate is that, for any curves J_1 and J_2, J_2 lies toward larger time than J_1,

(1.4.8) $$F(J_2) \le F(J_1),$$

provided that the total variation TV of the initial data is small and that M is chosen sufficiently large. By transcendental induction, it suffices to prove (1.4.8)

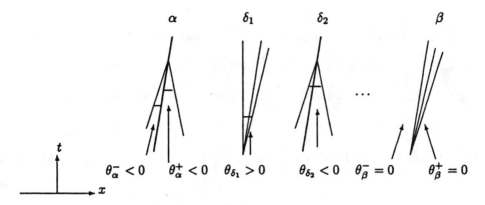

Figure 1.4.3. Effective angle between i-waves.

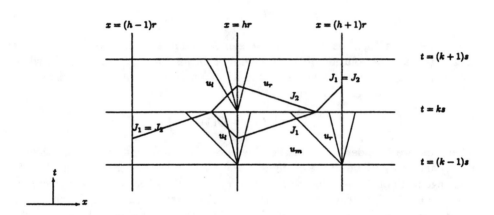

Figure 1.4.4. Local interactions.

when J_2 is an immediate successor of J_1, meaning that J_1 and J_2 differ only at one grid point; say, J_1 goes through $(hr, (k-1/2)s)$ while J_2 goes through $(hr, (k+1/2)s)$ and they sandwich a *diamond* $\triangle = \triangle_{h,k}$ with vertices $((h-1+a_k)r, ks)$, $((h+a_k)r, ks)$, $(hr, (k-1/2)s)$, and $(hr, (k+1/2)s)$. The waves entering $\triangle$ are part of the solutions of the Riemann problems issued from $(hr, (k-1)s)$ and either from $((h-1)r, (k-1)s)$ or $((h+1)r, (k-1)s)$, depending on whether $a_{k-1} \le 1/2$ or $a_{k-1} > 1/2$; see Figure 1.4.4. (Diamonds with vertices of the form $((h-1+a_k)r, ks)$, $((h-1+a_{k+1})r, (k+1)s)$, $((h-1)r, (k+1/2)s)$, and $(hr, (k+1/2)s)$ do not represent areas of wave interaction. Thus two J curves are regarded as the same if they differ only on such a diamond.)

The wave leaving $\triangle$ is the solution of the Riemann problem issued from (hr, ks). Thus the situation is the same as that dealt with in the last section when we studied the local wave interaction in Theorem 1.3.1. We denote by (u_l, u_m), (u_m, u_r) the

Riemann problems corresponding to the waves entering $\triangle$ and by (u_l, u_r) those that leave $\triangle$; see Figure 1.4.4. The amount of interaction within $\triangle$ is

$$(1.4.9) \qquad D(\triangle) \equiv D_d(\triangle) + D_s(\triangle) \equiv Q_d(u_l, u_m, u_r) + Q_s(u_l, u_m, u_r),$$

and, for later use, the amount of interaction in a region Λ is denoted by

$$(1.4.10) \qquad\qquad D(\Lambda) \equiv \sum \{D(\triangle_{i,j}) : (ir, js) \in \Lambda\}.$$

For the first curve J_0 between $t = 0$ and $t = s$, the functional is dominated by the total variation TV of the initial data:

$$F(J_0) = O(1)TV,$$

which is assumed to be small. To prove (1.4.8) by induction we assume

$$F(J_1) \le F(J_0) = O(1)TV.$$

The waves crossing J_1 and J_2 are the same outside $\triangle$ and, around $\triangle$, waves crossing J_1 are the solution of the Riemann problems (u_l, u_m) and (u_m, u_r), while those crossing J_2 are the solution of the Riemann problem (u_l, u_r). These waves are related according to Theorem 1.3.1; whence we have

$$L(J_2) \le L(J_1) + O(1)D(\triangle).$$

There are two considerations for the difference of the wave interaction functionals $Q(J_1)$ and $Q(J_2)$: Due to the changes in wave strengths after interaction, there is a change in the nonlinear functional of order $O(1)Q(\triangle)$ times the total strength, which is $O(1)TV$, of waves crossing the common part of J_1 and J_2. On the other hand, and this is the key point, waves entering $\triangle$ are interacting with the measure of interaction $D(\triangle)$, while those leaving $\triangle$ are the solution of a Riemann problem and are therefore noninteracting. For the quadratic wave interaction measure Q_d the above two considerations yield

$$Q_d(J_2) - Q_d(J_1) \le O(1)TV \cdot D(\triangle) - D_d(\triangle).$$

The cubic measure Q_s requires some computations. We consider Q_s^k when the two k-waves before interaction are shocks. Let δ_k be a k-wave to the right. Assume that these waves are interacting in the sense that the angle Θ between them is negative; see (1.4.7). Then the potential interaction measure between δ_k and the k-waves entering $\triangle$ is

$$|\alpha_k||\beta_k|(\sigma_1 - \sigma_2) + (|\alpha_k|(\sigma_1 - \sigma_2) + (|\alpha_k| + |\beta_k|)|\Theta(\beta_k, \delta_k)|)|\delta_k|.$$

From conservation laws and Theorem 1.3.1, this equals

$$|\alpha_k||\beta_k|(\sigma_1 - \sigma_2) + (|\gamma_k|\sigma - (|\alpha_k| + |\beta_k|)\sigma_2 + (|\alpha_k| + |\beta_k|)|\Theta(\beta_k, \delta_k)|)|\delta_k|$$
$$+ O(1)D(\triangle)|\delta_k|.$$

The interaction measure between δ_k and the k-waves γ_k leaving $\triangle$ is

$$|\gamma_k||\Theta(\gamma_k, \delta_k)||\delta_k| = |\gamma_k|(\sigma - \sigma_2 + |\Theta(\beta_k, \delta_k)|)|\delta_k|.$$

Since $|\gamma_k| = |\alpha_k| + |\beta_k| + O(1)D(\triangle)$, the difference between interaction measures after and before the interaction is

$$-|\alpha_k||\beta_k||\sigma_1 - \sigma_2| + O(1)D(\triangle)|\delta_k|.$$

With the above analysis, we have

$$Q(J_2) - Q(J_1) = D(\triangle)(-1 + O(1)TV) \le -\frac{1}{2}D(\triangle),$$

where the last estimate is due to the smallness of TV. We conclude from the above estimates that, for TV sufficiently small and M chosen suitably large,

(1.4.11) $$F(J_2) - F(J_1) \le \left(O(1) - \frac{M}{2}\right) D(\triangle) \le -D(\triangle);$$

whence we have (1.4.8). For later uses we have, by telegraphing this estimate over a region Λ bounded by two curves J_- and J_+,

(1.4.12) $$D(\Lambda) \le F(J_+) - F(J_-).$$

Since the amount of cancelation is no greater than the existing waves, we have

(1.4.13) $$C(\Lambda) \le F(J_-) + O(1)D(\Lambda) \le 2TV.$$

Convergence. From the boundedness of the approximate solutions, it follows easily from Helly's theorem that there exists a sequence of mesh sizes tending to zero such that the approximate solutions tend to a limit function $u_*(x,t)$. This is done first for rational times and then we use the fact that the approximate solutions are continuous in t in the $L_1(x)$ topology:

(1.4.14) $$\int_{-\infty}^{\infty} |u(x, t_2) - u(x, t_1)|dx = O(1)|t_2 - t_1|.$$

This is a consequence of the finite speed of propagation of the scheme by requiring r/s to be bounded and satisfy the CFL condition: Assume, for simplicity, that the characteristic speeds λ_i, $i = 1, \ldots, n$, are away from zero through a change of independent variables if necessary. For a fixed x the variation in t of $u(x,t), t_1 < t < t_2$, is due to waves crossing the interval. These waves come from those in $u(y, t_1)$, $x - L(t_2 - t_1) < y < x + L(t_2 - t_1)$, $L = r/s$, and their interaction. The amount of interaction is bounded by the total strength of these waves as a consequence of Theorem 1.3.1. In summary, we have

$$
\begin{aligned}
|u(x, t_2) \quad &- \quad u(x, t_1)| \\
&= \quad O(1)\text{variation}\{u(y, t_1) : \ x - L|t_2 - t_1| < y < x + L|t_2 - t_1|\}.
\end{aligned}
$$

The estimate (1.4.12) follows from a change of order of the integration of the above estimate:

$$\int_{-\infty}^{\infty} |\mathbf{u}(x,t_2) - \mathbf{u}(x,t_1)| dx = \int_{-\infty}^{\infty} O(1) dx \int_{x-L|t_2-t_1|}^{x+L|t_2-t_1|} |d\mathbf{u}(y,t_1)|$$

$$= \int_{-\infty}^{\infty} O(1) L(t_2 - t_1) |d\mathbf{u}(y,t_1)| = O(1) L(t_2 - t_1) TV.$$

The equicontinuity of the approximation solution in the $L_1(x)$-norm, (1.4.14), and the pointwise convergence for rational times yield easily the almost everywhere convergence of the approximate solutions, for some choice of mesh sizes tending to zero, to a limit function $\mathbf{u}_*(x,t)$ for $-\infty < x < \infty$, $t \geq 0$.

Consistency. As we have seen in the example of the propagation of a single shock, the above limit function $u_*(x,t)$ cannot be a weak solution of (4.1) for any choice of the random sequence. The error is accumulated at $t = ks, k = 0, 1, \dots$:

$$\int_{-\infty}^{\infty} \int_{0}^{\infty} (u\phi_t + f(u)\phi_x)(x,t) dx dt + \int_{-\infty}^{\infty} (u\phi)(x,0) dx$$

(1.4.15)
$$= \sum_{k=0}^{MN} \int_{-\infty}^{\infty} (u(x,ks+0) - u(x,ks-0))\phi(x,ks) dx.$$

Here $\phi(x,t)$ is the test function with compact support $\phi(x,t) = 0$, $t > T = MNs$. (The choice of the form MNs is for later convenience when we let $M, N \to \infty$ as $s \to 0$.) It can be shown that, for almost all choices of the random sequence a_k, $k = 1, 2, \dots$, the error (1.4.15) tends to zero. This is due to some form of cancelation of the integrals. We will address this issue in a more definite way in the next section when we show that the error tends to zero for any given equidistributed sequence. For now we only calculate the measure of consistency (1.4.15) for the simple example of one shock studied in the paragraph immediate before Definition 1.4.1. By our study of shock location, then, we know that the limiting function in this case is a weak solution if the random sequence is equidistributed. Denote by $x = x(k)\Delta x$ the location of the shock at time $t = k\Delta t$. We have

$$\int_{-\infty}^{\infty} (u(x,ks+0) - u(x,ks-0))\phi(x,ks) dx$$

$$= \begin{cases} \int_{x(k)r}^{x(k)r+\sigma s}(u_+ - u_-)\phi(x,ks) dx & \text{if } a_k rx > \sigma s, \\ \int_{x(k)r+\sigma s}^{(x(k)+1)r}(u_- - u_+)\phi(x,ks) dx & \text{if } a_k r < \sigma s. \end{cases}$$

If we simplify the situation by assuming that the test function is a constant ϕ_0, then (4.15) becomes, for the interval $I = (0, \sigma s/r)$,

$$\sum_{k=0}^{MN} \int_{-\infty}^{\infty} (u(x,ks+0) - u(x,ks-0))\phi(x,ks) dx$$

$$= \phi_0(u_+ - u_-)(A(MN,I)(r - \sigma s) - A(MN,I^c)\sigma s)$$

$$= \phi_0(u_+ - u_-)T(A(MN, I))\left(\frac{r}{s} - \sigma\right) - (MN - A(MN, I))\sigma)\frac{1}{MN}$$

$$= \left(\frac{A(MN, I)}{MN} - \sigma\frac{s}{r}\right)\frac{r}{s},$$

which tends to zero as $MN \to \infty$ when the random sequence is equidistributed; see Definition 1.4.1. To deal with the nonconstancy of the test function $\phi(x, t)$, we divide the time zone $0 \le t < T = MNs$ into small time zones $\Lambda_p = \{(x, t) : N(p-1)s \le t < Nps\}$, $p = 1, 2, \ldots, M$. The test function is close to a constant in each time zone. The closeness is of the order $O(1)LNs = O(1)LT/M$; L is the Lipschitz constant of $\phi(x, t)$ and tends to zero as $M \to \infty$. In each time zone the random sequence becomes increasingly equidistributed as N becomes large. Thus the above analysis applies and we have established consistency for the propagation of a single shock as $M, N \to \infty$.   □

Remark. The wave interaction measure Q_s, (1.4.6), is third order. This is basic to the study of N-waves and more general time asymptotic behavior of the solutions [15]. The difficult paper [6] makes use of the third-order wave interaction measure in the study of regularity and decay of solutions for 2×2 systems. There is an alternative definition of Q_s used in [5] that is quadratic:

$$\bar{Q}_s^i \equiv \sum\{|\alpha||\beta| : \ \alpha \text{ and } \beta \text{ interacting } i\text{-waves crossing } J\}.$$

The corresponding global estimate (1.4.12) is

(1.4.16) $\bar{D}(\Lambda) \le \bar{F}(J_1) - \bar{F}(J_2)$

with the definition of Q_s in (1.3.1) replaced by

$$\bar{Q}_s^i(\mathbf{u}_l, \mathbf{u}_m, \mathbf{u}_r) \equiv \{\alpha_i\beta_i : \text{ one or both of } (\mathbf{v}_{i-1}, \mathbf{v}_i)$$
$$\text{and } (\mathbf{w}_{i-1}, \mathbf{v}_i) \text{ are shock waves}\}.$$

1.5   Nonlinear Superposition

The above construction of solutions using elementary waves as building blocks and studying the evolution of waves undergoing nonlinear interactions is a form of *nonlinear superposition*. This approach is made more definite and explicit with the following idea of *wave tracing*. The aim of this idea is to devise a bookkeeping scheme of subdividing the elementary waves so that the evolution of each subwave can be studied more definitely.

We explain the idea first for the scalar equation. Take the example of a solution $u(x, t)$ of two shocks (u_1, u_2), (u_2, u_3), $u_1 > u_2 > u_3$, of speeds σ_1, σ_2, respectively, combining into a single shock (u_1, u_3) of speed σ_3. We divide (u_1, u_3) into the superposition of the original two shocks. The result of the interaction is then viewed as if both of the original shocks keep their identities but with a change of speed; see Figure 1.5.1.

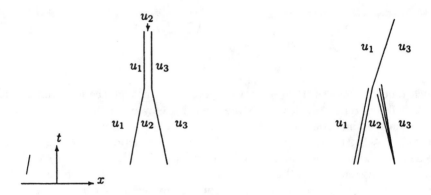

Figure 1.5.1. Wave combining.

Notice that the change of the wave speed $[\sigma]$ times the wave strength is

$$|[\sigma]_1||u_1 - u_2| + |[\sigma]_2||u_2 - u_3| = O(1)|u_1 - u_2||u_2 - u_3| = O(1)\bar{Q}_s(u_1, u_2, u_3),$$
$$[\sigma]_1 \equiv \sigma_1 - \sigma_3, \qquad [\sigma]_2 \equiv \sigma_2 - \sigma_3.$$

Here we have noticed that

$$|\sigma_1 - \sigma_3| = O(1)|u_2 - u_3|, \qquad |\sigma_2 - \sigma_3| = O(1)|u_1 - u_2|$$

due to the continuous dependence of a shock's speed on its end states.

Consider next the cancelation of a rarefaction (u_1, u_2) and a stronger shock (u_2, u_3), $u_2 > u_1 > u_3$, with speed σ_1. After the interaction, the rarefaction wave is canceled, so a portion of the shock is also canceled. We divide the shock (u_2, u_3) into subshocks (u_2, u_1) and (u_1, u_3); see Figure 1.5.1. The nonlinear interaction is then viewed as the rarefaction wave (u_1, u_2) and the subshock (u_2, u_1) canceling each other, while the subshock (u_1, u_3) with speed σ_2 survives. Denote by $C(u_1, u_2, u_3) \equiv |u_1 - u_2|$ the amount of wave cancelation. The change $[\sigma]$ of the surviving shock's speed is then

$$[\sigma] \equiv |\sigma_1 - \sigma_2| = O(1)C(u_1, u_2, u_3).$$

We may perform this partition of waves in an approximate solution in such a way that we obtain a mechanism for *wave tracing*. This is done as follows: Fix a small time $t_1 = N\Delta t$ and consider the approximate solution $u(x, t) = u_1 r(x, t)$ in the time zone $0 < t < t_1$. Waves interact and cancel in the time zone in a way that is not easily foreseen because of the nonlinearity and the randomness of the scheme. The *wave partition* is an a posteriori bookkeeping scheme. Given a shock at time $t = 0$ we partition it into subshocks sufficiently fine that each subshock is either canceled completely or survives intact in the zone. The situation is more complicated for a rarefaction wave: In addition to the cancelation, a rarefaction wave is divided when the random number a_k times Δx equals Δt times one of the

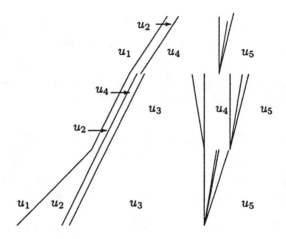

Figure 1.5.2. Wave combining and canceling.

characteristic speeds of the rarefaction wave. Nevertheless we may keep refining a partition of a rarefaction wave so that each subwave is either completely canceled or propagates intact as a single wave.

Notice that how fine a given wave needs to be partitioned and which subwaves survive depend on the random sequence as well as the time zone. This is expected as the waves behave nonlinearly. In Figure 1.5.2, two shock waves (u_1, u_2), (u_2, u_3) combine and then cancel with part (u_3, u_4), $u_2 > u_4 > u_3$, of a rarefaction wave (u_3, u_5). In this case the combined shocks (u_1, u_3) are partitioned into three waves. The shock (u_1, u_2), the subshock (u_2, u_4) of (u_2, u_3), and the rarefaction wave (u_4, u_5) are the surviving waves during the depicted time period.

The shock (u_1, u_2) changes its speed by the amount of $[\sigma] = O(1)(|u_2 - u_3| + |u_3 - u_4|)$ and so

$$|u_1 - u_2|[\sigma] = O(1)(\bar{Q}_s(u_1, u_2, u_3) + |u_1 - u_2|C(u_1, u_3, u_4)).$$

Next we turn to the system. In addition to wave combining and canceling, wave interaction may alter the wave states and produce new waves. Nevertheless, the above partitioning of waves can be generalized and we have three categories of waves: surviving ones, canceled ones, and those produced by interactions. We summarize the result in the following theorem.

THEOREM 1.5.1. *The waves in an approximate solution in a given time zone* $\Lambda = \{(x, t) : -\infty < x < \infty,\ k_1 s \leq t < k_2 s\}$ *can be partitioned into subwaves of categories* I, II, *or* III *with the following properties:*

(i) *The subwaves in* I *are surviving. Given a subwave* $\alpha(t)$, $k_1 s \leq t < k_2 s$, *in* I, *write* $\alpha \equiv \alpha(k_1 s)$ *and denote by* $|\alpha(t)|$ *its strength at time t, by* $[\sigma(\alpha)]$ *the variation of its speed, and by* $[\alpha]$ *the variation of the jump of the states across it over the* *time interval $k_1 s \le t < k_2 s$. Then*

$$\sum_{\alpha \in I} ([\alpha] + |\alpha(k_1 s)|[\sigma(\alpha)]) = O(1)\bar{D}(\Lambda).$$

(ii) *A subwave $\alpha(t)$ has positive initial strength $|\alpha(k_1 s)| > 0$ but it is canceled in the zone Λ, $|\alpha(k_2 s)| = 0$. Moreover, the total strength and variation of the wave shape satisfies*

$$\sum_{\alpha \in II} ([\alpha] + |\alpha(k_1 s)|) = C(\Lambda) + O(1)\bar{D}(\Lambda).$$

(iii) *A subwave in* III *has zero initial strength $|\alpha(k_1 s)| = 0$ and is created in the zone Λ, $|\alpha(k_2 s)| > 0$. Moreover, the total variation satisfies*

$$\sum_{\alpha \in III} ([\alpha] + |\alpha(t)|) = O(1)\bar{D}(\Lambda), \qquad k_1 s \le t < k_2 s.$$

As an application of the theorem, we have the following deterministic version of the Glimm scheme in Theorem 1.5.2.

THEOREM 1.5.2. *Suppose that the random sequence a_k, $k = 1, 2, \dots$, is equidistributed. Then the limit function $\mathbf{u}_*(x, t)$ of the Glimm scheme is a weak solution of the hyperbolic conservation laws.*

Proof. With the wave properly partitioned, we may trace the surviving waves with sufficient accuracy as follows: For the scalar equation, a rarefaction wave propagates with constant speeds and so its location can be predicted by the scheme (1.4.5). The error in (1.4.13) due to a subrarefaction wave α is $|\alpha|$ times a quantity, which tends to zero as $M, N \to \infty$. This follows from the same analysis we carried out for a single shock toward the end of the last section. On the other hand, a subshock α has a varying speed and the analysis does not apply directly. The scheme (1.4.5) can determine its location only up to the variation of its speed in the time zone. The variation of the speed is of the order of wave combining and cancelation that happens to the shock of which the subshock is a part. From the analysis of wave interaction in Theorem 1.3.1, we know that $|\alpha|$ times the variation of its speed is of the order of the measure of wave interaction $\bar{D}(\Lambda)$. This is so when we sum up all the subshocks; see part (i) of Theorem 1.5.1. Thus the new error contributed by subshocks to the measure of consistency in a given time zone $Z_l \equiv \{(x, t) : x \in R, (l-1)Ns \le t < lNs\}$ is $O(1)\bar{D}(Z_l)Ns = O(1)D(Z_l)T/M$. The total new error over $0 \le t < T$ is then $O(1)\bar{D}(t \ge 0)T/M$, which tends to zero as $M \to \infty$. The error contributed by the canceled subwaves in a time zone Z_l is $O(1)C(Z_l)T/M$ and is dealt with similarly. Thus the total error is of the form $O(1)[(A(N, I)/N - |I|) + (D + C)(t \ge 0) \cdot T/M]$, which tends to zero as $M, N \to \infty$.

Notice that in the above we have made use of the boundedness of the total cancelations and interactions in $\{(x, t) : x \in R, t \ge 0\}$, (1.4.11), and also the fact that the wave partition is done independently for each time zone Z_l, $l = 1, 2, \dots, M$, so that the variation of the speed of each subshock is $\bar{Q}(Z_l)$. A one-time partition for the entire region $\{(x, t) : x \in R, t \ge 0\}$ would be too crude and does not yield the vanishing factor T/M in the above error estimate.

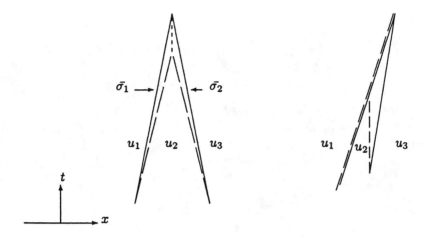

Figure 1.5.3. Linear superposition.

We now turn to the system. There are two new elements we need to deal with: the changes in the jumps of the wave states of surviving waves and the waves produced due to interactions. Either one is controlled by the interaction measure $\bar{D}(\Lambda)$. The situation is therefore similar to the case of the scalar law above and we omit the details. □

Another application of wave tracing is that it is useful for the study of the evolution of the $L_1(x)$-norm. We will show that the approximate solutions, and thereby the exact solution, can be approximated locally in time with a wave pattern $\bar{u}(x,t)$ of *linear superposition of nonlinear waves* constructed as follows: We assume that the solution minus its limit at $x = \pm\infty$ is in $L_1(x)$. This allows us to ignore, within any degree of accuracy in $L_1(x)$, the waves near $x = \pm\infty$ and to consider only a finite number of subwaves in an approximate solution in a given time zone $\Lambda \equiv \{(x,t) : -\infty < x < \infty, \ 0 \leq t < T = MNs\}$. We number the surviving i-waves by $\alpha_i^1(t), \alpha_i^2, \ldots, \alpha_i^K$, located from left to right at $x = x_1(t) \leq x_2(t) \leq \cdots \leq x_K(t)$. For each i-wave α_i^k we construct an approximate i-wave $\bar{\alpha}_i^k$ with the same states as $\alpha_i^k(0)$ at time $t = 0$ and propagate along the straight line joining the end positions of the wave $(x_k(0), 0)$ and $(x_k(T), T)$. Any nonsurviving wave α at time $t = 0$ is assigned the corresponding wave $\bar{\alpha}$ with the same states. However, since α does not exist at time $t = T$, and therefore has no location then, we simply assign its end location at time $t = T$ in such a way that the wave $\bar{\alpha}$ propagates with constant speed closest to the initial speed of α, and no two i-waves intersect in the zone Λ. The wave pattern $\bar{u}(x,t)$ consists of these approximate nonlinear waves $\bar{\alpha}_i^k$ propagating along straight lines.

THEOREM 1.5.3. *The wave pattern $\bar{u}(x,t)$ is close to the approximate solution*

$\mathbf{u}_r(x,t)$ *in* $L_1(x)$*-norm:*

(1.5.1) $\int_{-\infty}^{\infty} |\mathbf{u}_r(x,T) - \bar{\mathbf{u}}(x,T)| dx = O(1)((D+C)(\Lambda) + B(N)TV)T,$
             $\Lambda \equiv \{(x,t) : \; -\infty < x < \infty, \; 0 \le t < T\}.$

Here $B(N)$ *is the* sup *of the measure* $B(N, I)$ *of the equidistributedness of the random sequence over all subintervals* I *of* $(0, 1)$.

Proof. Take the simple example of the *linear superposition* $\bar{u}(x,t)$ of the two shocks for the scalar equation (see Figure 1.5.3); cf. Figure 1.5.1.

The time change of their L_1-distance is

$$\frac{dt}{d} \int |u(x,t) - u_*(x,t)| dx \;\; = \;\; |\sigma_1 - \sigma_3||\bar{v}| + |\sigma_2 - \sigma_3||\bar{w}|$$
$$= \;\; O(1)|\bar{v}||\bar{w}| = O(1)\bar{Q}(u_1, u_2, u_3).$$

Here we have noticed that $\sigma_2 - \sigma_3 = O(1)|\bar{v}|$ and that $\sigma_1 - \sigma_2 = O(1)|\bar{w}|$ by the continuous dependence of a shock's speed on its end states and that the speed $\bar{\sigma}_1$ (or $\bar{\sigma}_2$) of the shocks (u_1, u_2) (or (u_2, u_3)) in the wave pattern $\bar{u}(x,t)$ lies between σ_1 (or σ_2) and σ_3. Consider next the cancelation of a rarefaction (u_1, u_2) and a stronger shock (u_2, u_3), $u_2 > u_1 > u_3$; cf. Figure 1.5.1. The time change of the $L_1(x)$ distance between the approximate solution with the linear superposition of the shock and rarefaction wave, Figure 1.5.3, is $O(1)(C + |u_1 - u_2||u_2 - u_3|)$. The rest of the proof follows that of the last theorem and is omitted.   □

We note that the theorem is different from the $L_1(x)$ Lipschitz continuity of the solutions as stated in the existence theorem of the last section. Here the Lipschitz constant $O(1)(\bar{D} + C)(\Lambda)$ does not accumulate; instead, it sums up to a finite constant $O(1)(\bar{D} + C)(\{(x,t) : -\infty < x < \infty, \; t \ge 0\}$. In other words, as far as $L_1(x)$ topology is concerned, and subject to the global estimates on wave interactions, the solution is approximated, uniformly in time, by the wave pattern of linear superposition of nonlinear waves. There has been important progress toward the well-posedness theory for a system of hyperbolic conservation laws. The right norm is $L_1(x)$. The $L_2(x)$-norm is convenient for a functional analytic approach. However, the norm is not suitable for continuous dependence on the initial data, because the $L_2(x)$ distance between dissipative rarefaction waves decays, but between antidissipative compression waves it can increase at an arbitrary rate. With the above theorem, the study of the $L_1(x)$ well-posedness theory is then not about the nonlinear wave interactions, which have been considered in the Glimm scheme and the subsequent analysis explained so far. Instead, it is about the coupling of different wave patterns, each consisting of a linear superposition of nonlinear waves. This is not to say that the Glimm analysis is not needed, but the additional efforts are orthogonal to it. The goal is to find a nonlinear functional that is equivalent to the L_1-distance between the solutions and is time decreasing. This has been carried out in [19] and [20]. There is another, different recent approach on the well-posedness problem initiated by Bressan, Goatin, and Piccoli based on a detailed study of the interaction of waves [1], through a wave tracking approximate method [3].

The wave tracing is a convenient mechanism for defining the generalized characteristics, a notion that has been shown to be useful for the study of the behavior of solutions for the scalar equation; see section 1.1. For any surviving wave α in the partition, Theorem 1.5.1, its location defines a generalized characteristic curve. Given two i-characteristics Ξ_i^1 and Ξ_i^2, consider the region Ω_i between them and time t_1 and time t_2. The following *conservation of waves* is an consequence of the fact that an i-wave does not cross an i-characteristic:

$$(1.5.2) \qquad \begin{aligned} X_i^+(t_2) &= X_i^+(t_1) - C(\Omega_i) + O(1)\bar{D}(\Omega_i), \\ X_i^-(t_2) &\leq X_i^-(t_1) - C(\Omega_i) + O(1)\bar{D}(\Omega_i). \end{aligned}$$

Here X_i^+ (or X_i^-) denotes the number of i-rarefaction (or shock) waves in the region.

1.6   Large-Time Behavior and Regularity

We show that, due to nonlinearity, the solution obtained in the last section has a simple large-time state, the solution of the Riemann problem

$$(1.6.1) \qquad (\mathbf{u}_l, \mathbf{u}_r), \quad \mathbf{u}_l \equiv \mathbf{u}_0(-\infty), \quad \mathbf{u}_r \equiv \mathbf{u}_0(\infty).$$

Thus the solution, time asymptotically, depends only on the end states of the initial data $\mathbf{u}(x,0) = \mathbf{u}_0(x)$. There is a necessary shift of the location of the elementary waves in the solution of the Riemann problem, of course. Our approach differs from that toward the end of section 1.1.1, which yields the rate of convergence to the time-asymptotic states, since it applies to a more general situation without, for instance, the genuine nonlinearity assumption. (For simplicity, we assume that the system is genuinely nonlinear [15].)

THEOREM 1.6.1. *Suppose that system* (1.1) *is genuinely nonlinear. Then the weak solution obtained in Theorem* 1.4.1 *approaches the wave pattern* $(\mathbf{u}_{i-1}, \mathbf{u}_i)$, $i = 1, 2, \ldots, n$ *in the solution of the Riemann problem* (1.6.1) *as time tends to infinity.*

Proof. The key observation is that as time increases the solution becomes less interactive as a consequence of the boundedness of the total number of actual interactions (1.4.16). We will use Proposition 1.2.1 also for our analysis. This is made precise as follows: From (1.4.16) we know that the total amount of actual interaction $D(\Lambda)$ in any given region Λ is finite. Consequently, given any $\delta > 0$ there exists time T_0 such that the total amount of actual interaction after time T_0 is less than δ^2: $(D+C)(\Lambda) < \delta^2$, $\Lambda \equiv \{(x,t) : -\infty < x < \infty, t > T\}$. (Notice that the measures Q and C have been shown to be uniformly bounded in the Glimm estimates. Thus we may assume that they converge in weak* measure as the approximate solutions converge to the exact solution. Thus the time T_0 is chosen to have the desired property also for approximate solutions with sufficiently small meshes r, s. This allows us to construct the generalized characteristics for the approximate solutions by the wave tracing method and obtain estimates uniform in mesh sizes.) We will apply Proposition 1.2.1 as follows.

Since the solution is of bounded variation, we may choose M sufficiently large so that the variation of $\mathbf{u}(x,t)$ over $|x| > M$ is less than δ. Through $x = \pm M$

draw generalized characteristics $\Xi_i^\pm$, $i = 1, 2, \ldots, n$. These curves of different characteristic families intersect before time $T = T_0 + O(1)M$. Denote the region between Ξ_i^- and Ξ_i^+ after time T by Ω_i. The j-waves, $j \neq i$, in Ω_i come either from the $|x| > M$ at time T or are produced by interactions and are therefore of total strength $O(1)\delta^2$; cf. (1.4.16) and (1.5.2). By the continuous dependence of the Riemann solution on its data, to prove the theorem it suffices to show that the i-waves in Ω_i are dominated either by a single i-shock or by the i-rarefaction waves. By the same reasoning, we may assume, for simplicity, that the wave pattern $\mathbf{W}_i$ in Ω_i at time $T + O(1)M$ consists of i-waves only. We know that the total amount of actual wave interaction and cancelation between these i-waves after that time is less than δ^2.

There are two cases: In the first case, the wave pattern $\mathbf{W}_i$ is divergent up to $O(1)\delta$ in the sense of Proposition 1.2.1 and is therefore dominated by the i-rarefaction waves. Through each point $(x, t) \in \Omega_i$, $t > T$, draw an i-generalized characteristic X^i backward in time. We do this for a given approximate solution according to the wave tracing as follows: If there is a shock through (x, t), we take the characteristic speed of either side of the shock to propagate, say, the right side. At each time step $t = ks$, X^i moves to the grid point according to the random sequence a_k. We take the closest i-characteristic issued from the grid point to follow backward in time. The variation of the speed is due to wave crossing and cancelation. The total amount of the former is less than the total amount of j-waves, $j \neq i$, in Ω_i. Thus they are of the order $O(1)\delta^2$. The curve X^i is not continuous; it jumps at each time step. However, this error is taken care of by the equidistributedness of the random sequence. Thus we conclude that, in the limit of zero mesh sizes, the curve Ξ^i is of speed $\lambda_i(\mathbf{u}(x, t)) + O(1)\delta^2$. It lands in $|x| < M$ at time T. Thus

$$x = O(1)M + (\lambda_i(\mathbf{u}(x, t)) + O(1)\delta)(t - T) \text{ or}$$
$$\lambda_i(\mathbf{u}(x, t)) - \frac{x}{t} = O(1)\frac{M}{t} + O(1)\delta + O(1)(\lambda_i(\mathbf{u}(x, t)) + \delta)\frac{T}{t}.$$

Since we already know that the solution takes values in a small neighborhood of a rarefaction curve in Ω_i, it approaches, within the arbitrarily small error $O(1)\delta$, the centered rarefaction wave there:

$$\lambda_i(\mathbf{u}(x, t)) - \frac{x}{t} \to O(1)\delta.$$

In the second case, there exist locations $x_1 < x_2$ such that $\lambda_i(x_1) > \lambda_i(x_2) + \delta$. Through (x_i, T), $i = 1, 2$, we draw generalized characteristics. These curves propagate with speed $\lambda_i(x_j) + O(1)\delta^2$, and their speed difference is less than

$$-\delta + O(1)\delta^2 < -\frac{1}{2}\delta,$$

and so they must meet in finite time to form a shock. In fact, the waves next to the shock will interact with the shock in finite time for the same reason. In other words, the i-waves are dominated by a single shock. This completes the proof of the theorem. □

Notice in the above proof that the approximate solutions tend to the time-asymptotic states at the same rate as the exact solutions. In other words, the deterministic version of the Glimm scheme is accurate time asymptotically.

We now turn to the study of the regularity of the solutions. The interaction measure D is finite and so can have positive point measure only at countable points. We will show that, except for these countable points of wave interaction, the solution consists of countable shock curves and the rest are continuity points. Because this is the same as in the above time-asymptotic study, we will see that this regularity property is already in place for the approximate solutions.

THEOREM 1.6.2. *The physical space $\{(x,t) : -\infty < x < \infty, \ t \geq 0\}$ consists of countable points P of wave interactions, countable Lipschitz shock curves S, and the continuity points C for the solution $\mathbf{u}(x,t)$ constructed by the Glimm scheme. More precisely, we have the following:*

(i) In a small neighborhood of $(x_0, t_0) \in P$, the solution is dominated by the Riemann solution $(\mathbf{u}_l, \mathbf{u}_r)$, $\mathbf{u}_l \equiv \mathbf{u}(x_0 - 0, t_0)$, $\mathbf{u}_r \equiv \mathbf{u}(x_0 + 0, t_0)$, in the forward time $t > t_0$.

(ii) In a small neighborhood of $(x_0, t_0) \in S$, the solution is dominated by a single shock satisfying the jump and entropy conditions. Moreover, for sufficiently small mesh sizes r, s, there is a dominating shock in the approximate solutions that tends to the shock in the exact solution.

(iii) At any point (x_0, t_0) not in P and S, the solution is continuous. Moreover, for any given $\delta > 0$, there exists a small neighborhood of (x_0, t_0) with the property that the approximate solutions there have oscillations less than δ for the mesh sizes sufficiently small. In other words, the approximate solutions tend to the exact solution, not only pointwise but locally at (x_0, t_0).

Proof. Consider first a point (x_0, t_0) at which there is no nonzero point measure of $D + C$. We have two cases: $\mathbf{u}_l \neq \mathbf{u}_r$ and $\mathbf{u}_l = \mathbf{u}_r$, $\mathbf{u}_l \equiv \mathbf{u}(x_0 - 0, t_0)$, $\mathbf{u}_r \equiv \mathbf{u}(x_0 + 0, t_0)$. The first case corresponds to a shock; the second case corresponds to the continuity point. Consider the continuity point first. Given a small $\delta > 0$, there exists a neighborhood $\Omega \equiv \{(x,t) : (x - x_0)^2 + (t - t_0)^2 < c^2\}$ of (x_0, t_0) with the property that $(D+C)(\Omega) < \delta^2$ and that the total variation in x of $\mathbf{u}(x, t_0)$, $(x, t_0) \in \Omega$, is less than δ. The former holds for sufficiently small mesh sizes, because there is no point measure of $D+C$ at (x_0, t_0) and these measures converge in the weak* sense. The latter holds because the solution $\mathbf{u}(x,t)$ is of bounded total variation in x and is continuous in x at (x_0, t_0). However, it is not immediately clear that this is also true for the approximate solutions because there may be cancelations in the limit, although not in the approximate solutions. (The cancelations in the approximate solutions in Ω are assumed to be small.) Nevertheless, because of pointwise almost everywhere convergence proved in Theorem 1.5.2, there exist $x_\pm$, $x_0 - 2\delta^2 < x_- < x_0 - \delta^2$, $x_0 + \delta^2 < x_- < x_0 + 2\delta^2$, such that, for the mesh sizes sufficiently small, the approximate solutions satisfy $|\mathbf{u}(x_-, t_0) - \mathbf{u}(x_+, t_0)| < 2\delta$. We claim that, except for an amount no larger than δ, waves crossing the interval $(x_0 - 2\delta^2, x_0 + 2\delta^2)$ belong to a particular, say, ith characteristic family. This is so because otherwise waves of different families would interact in Ω and make $D(\Omega) > \delta^2$. Next we claim that there are no compression i-waves of strength, measured in the decrease in λ_i, greater than $\sqrt{\delta}$ crossing the interval $(x_0 - 2\delta^2, x_0 + 2\delta^2)$ at time t_0. Otherwise, they would form a shock in Ω of strength greater than $\sqrt{\delta}$ and the shock would in turn attract the waves crossing the interval. Since we have $|\mathbf{u}(x_-, t_0) - \mathbf{u}(x_+, t_0)| < 2\delta$, this would imply that there is a cancelation of the order $\sqrt{\delta} + O(1)\delta > \delta$ in Ω, contradicting the hypothesis $C(\Omega) < \delta$. Thus we conclude that there is no strong compression wave and so Proposition 1.2.1 applies and we conclude that $|\mathbf{u}(x, t_0) - \mathbf{u}(x_0, t_0)| = O(1)\delta$, $x \in (x_0 - 2\delta^2, x_0 + 2\delta^2)$. To check the oscillation of the approximate solutions for time near t_0, we trace the waves to first verify that $|\mathbf{u}(x_-(t), t) - \mathbf{u}(x_+(t), t)| < 2\delta$ for time t close to t_0 and along the generalized characteristics $x = x_\pm(t)$ through $(x_\pm, t_0)$. Then we apply Proposition 1.2.1 as above. This completes the proof of (iii) of the theorem.

Next we consider a point (x_0, t_0) at which there is no nonzero point measure of $D + C$ and where $\mathbf{u}_l \neq \mathbf{u}_r$, $\mathbf{u}_l \equiv \mathbf{u}(x_0 - 0, t_0)$, $\mathbf{u}_r \equiv \mathbf{u}(x_0 + 0, t_0)$. We choose δ to be $|\mathbf{u}_r - \mathbf{u}_l|^2$. Again we choose a small neighborhood Ω of (x_0, t_0) with $(D+C)(\Omega) < \delta^2$. As in the above analysis using (1.5.2), most of the waves crossing the interval $(x_0 - 2\delta^2, x_0 + 2\delta^2)$ at time t_0 belong to a particular ith characteristic family. Similarly we have, for approximate solutions with sufficiently small mesh sizes, $|\mathbf{u}(x_-, t_0) - \mathbf{u}_l| + |\mathbf{u}(x_+, t_0) - \mathbf{u}_r| < \delta$. By the same reasoning as above, we have either a relatively strong i-shock or i-rarefaction waves. The i-shock cannot be formed by compression in Ω, which would create an interaction measure of order greater than δ^2. Thus a single i-shock dominates the solution in Ω and (ii) is proved. In the case where i-rarefaction waves dominate, most of them have to be surviving waves in wave tracing because $(D + C)(\Omega)$ is assumed to be small. On the other hand, the i-rarefaction becomes compressed in the backward time direction and cannot be traced backward in time by simple geometric reasoning. Thus i-rarefaction waves cannot dominate in the region.

Finally we consider an interaction point (x_0, t_0). Choose two neighborhoods of (x_0, t_0), the larger one Ω_b with radius δ and the smaller one Ω_s with radius δ^2 such that, for sufficiently small mesh sizes, $(D+C)(\Omega_s) > \delta$, $(D+C)(\Omega_b - \Omega_s) < \delta^2$. Through $(x_0 \pm \delta^2, t_0)$ draw generalized characteristics $X_i^\pm$, $i = 1, 2, \ldots, n$. Between X_i^- and X_i^+, inside Ω_b and outside Ω_s, i-waves dominate. Moreover, in the same region and for time greater than t_0, there are no strong compression waves because the distance between X_i^- and X_i^+ is $O(1)\delta^2$ when they just leave Ω_s (at time t_1, say) and this leaves little room for compression. Indeed, if there are compression wave of strength $\lambda_i(x_1, t_1) - \lambda_i(x_2, t_1) > \delta$, $x_1 < x_2$, then the generalized characteristics through (x_1, t_1) and (x_2, t_1) will intersect in Ω_b to form a shock. This shock must exist already at time t_1; otherwise, the compression would create interactions of an amount greater than δ^2 in Ω_b. Thus we conclude that in the region between X_i^- and X_i^+, inside Ω_b and outside Ω_s, either a single i-shock dominates or i-rarefaction waves dominate. As the mesh sizes tend to zero, we may let the region Ω_s converge to the point (x_0, t_0). This proves statement (i) of the theorem. $\square$

CHAPTER 2

Viscous Conservation Laws

2.1 Preliminaries

In many physical situations, dissipations, such as viscosities and heat conduction in the gas dynamics, are important. The simplest way to incorporate this is to add higher-order terms to the hyperbolic conservation laws to yield the *viscous conservation laws*, which usually take the form

(2.1.1) $$\mathbf{u}_t + f(\mathbf{u})_x = (B(\mathbf{u}, \varepsilon)\mathbf{u}_x)_x.$$

Here the *viscosity matrix* $B(\mathbf{u}, \varepsilon)$ also depends on the *dissipation parameters* ε. The vector ε has nonnegative components and as it tends to zero the system becomes the hyperbolic conservation laws

$$B(\mathbf{u}, 0) = 0,$$
(2.1.2) $$\mathbf{u}_t + f(\mathbf{u})_x = 0.$$

In the theory for hyperbolic conservation laws we need to impose the entropy condition on the shocks for well-posedness. The entropy condition can be captured by considering the hyperbolic conservation laws as the *zero dissipation limit*, $\varepsilon \to 0$, of the viscous conservation laws. We now show this for the scalar conservation law: The entropy condition (1.1.9) of Chapter 1 for convex scalar laws has been well motivated as compression. We now consider the scalar law $u \in \mathbb{R}^1$ when the flux function $f(u)$ is not necessarily convex and may have inflection points. In this case we have the *Oleinik entropy condition*

(2.1.3) $$s = \frac{f(u_+) - f(u_-)}{u_+ - u_-} \le \frac{f(u) - f(u_-)}{u - u_-}.$$

The condition reduces to the compressibility condition (1.1.9) of Chapter 1 when the flux satisfies $f''(u) \ne 0$. For a general flux it implies the weak compressibility

$$f'(u_+) \le s \le f'(u_-).$$

To justify the entropy condition (2.1.3), we consider the viscous conservation law

(2.1.4) $$u_t + f(u) = \varepsilon u_{xx}$$

39

and traveling waves corresponding to the inviscid shock (u_-, u_+):

(2.1.5) $\begin{aligned} &u(x,t) = \phi(\xi), \quad \xi \equiv \frac{x-st}{\varepsilon}, \quad \phi(\pm\infty) = u_\pm, \\ &\phi'(\xi) = f(\phi) - f(u_-) - s(\phi - u_-) \equiv g(\phi), \quad g(u_\pm) = 0. \end{aligned}$

There are two points to this exercise: The viscous conservation law is more complete from the physical point of view due to the inclusion of the viscous effects on the right-hand side of (2.2.1). This makes the system parabolic and it thus possesses only smooth solutions and so there is no need for an entropy condition. Then, as the viscosity ε tends to zero, the traveling wave $\phi(\xi)$, the *viscous shock*, tends to the discontinuous hyperbolic shock (u_-, u_+) of (2.1.2). This is because the viscous shock $\phi(\xi)$ is a function of $\xi = (x - st)/\varepsilon$, which implies that the shock layer is of width ε. We can now state the following admissibility criterion, formulated for a general system of conservation laws.

DEFINITION 2.1.1. *A shock wave $(\mathbf{u}_-, \mathbf{u}_+)$ with speed s for the hyperbolic conservation laws* (2.1.2) *is called admissible according to the* viscosity criterion *with respect to the viscosity matrix $B(\varepsilon, \mathbf{u})$ if there exists a traveling wave of the viscous conservation laws* (2.1.1) *with speed s and end states $\mathbf{u}_\pm$ at $x = \pm\infty$.*

The mathematical requirements on the viscosity matrix $B(\mathbf{u}, \varepsilon)$, beyond that (2.1.1) reduces to (2.1.2) and $B(\mathbf{u}, 0) = 0$, are motivated by physical models and will be discussed later. For the scalar conservation law, any positive viscosity gives rise to the same entropy condition. For simplicity, we have taken $B(u, \varepsilon) = \varepsilon$ in (2.1.4), the *artificial viscosity*. The ODE (2.1.5) has a solution if and only if the function $g(\phi)$ is positive (or negative) for $u_- < \phi < u_+$ (or $u_- > \phi > u_+$). It is easy to see that this is equivalent to

(2.1.6) $s = \dfrac{f(u_+) - f(u_-)}{u_+ - u_-} < \dfrac{f(u) - f(u_-)}{u - u_-}$ for any u between u_- and u_+.

One may combine several limits of viscous shocks with the same speed to form a single hyperbolic shock and obtain the original Oleinik entropy condition (2.1.3). Thus we have the following proposition.

PROPOSITION 2.1.1. *The viscosity criterion and the Oleinik entropy condition are equivalent for the scalar conservation law.*

The entropy condition for a system of hyperbolic conservation laws, Definition 1.2.1, (1.2.10) of Chapter 1, can also be justified the same way according to the viscosity criterion, Definition 2.1.1, [24]. This is so for strictly hyperbolic systems. There are richer wave phenomena for the viscous conservation laws when the associated inviscid system is nonstrictly hyperbolic; see [4], [22], and references therein.

As with the hyperbolic systems, the main concern with the viscous systems is the coupling of waves pertaining to different characteristic families. There is the added complexity, already present for the scalar equation, resulting from the combined effect of nonlinearity and dissipation. The situation is therefore richer and more complex than for the hyperbolic conservation laws. So far, these issues have been studied with regard to the stability of nonlinear waves. In the next section we consider the scalar equation, the Burgers equation. The equation is endowed with a scaling property and has explicit solutions. We will make use of these to illustrate the common properties of convex scalar conservation laws.

The study of nonlinear waves for systems is illustrated in the subsequent sections. We start with the simple situation of the perturbation of a constant state in section 3, which reveals some of the interesting effects of the coupling of waves of different characteristic families. The study of shock and rarefaction waves is done in sections 4 and 5. There is a major difference between scalar and systems of viscous conservation laws in that the viscosity matrix $B(u,\varepsilon)$ is not positive definite for general physical systems. Thus (2.1.1) is not uniformly parabolic but *hyperbolic parabolic*. This has important implications for the smoothing property of the solution operator [9], [21]. We will, however, deal only with systems with artificial viscosity in the remainder of this chapter.

2.2 The Burgers Equation

To gain some preliminary understanding of the interplay of dissipation and nonlinearity of the flux function, we consider the *Burgers equation*

$$(2.2.1) \qquad u_t + uu_x = \varepsilon u_{xx}.$$

This basic equation has a certain scaling property and can be solved explicitly by transforming it into the heat equation through the *Hopf–Cole transformation* [27]:

$$u \equiv -2\varepsilon \frac{w_x}{w},$$

$$(2.2.2) \qquad w_t = \varepsilon w_{xx}.$$

The initial value problem for the heat equation can be solved through the heat kernel

$$w(x,t) = \int_{-\infty}^{\infty} (4\pi\varepsilon t)^{-1/2} e^{\frac{(x-y)^2}{4\varepsilon t}} \, dy.$$

Thus the solution of the initial value problem for the Burgers equation (2.2.1) is

$$u(x,t) = \frac{\int_{-\infty}^{\infty} \frac{x-y}{t} e^{-U/2\varepsilon} dy}{\int_{-\infty}^{\infty} e^{-U/2\varepsilon} dy},$$

$$(2.2.3) \qquad U = U(x,t;y) \equiv \frac{(x-y)^2}{2t} + \int_0^y u(z,0)dz.$$

From this explicit formula much is learned about the behavior of the solutions of the Burgers equation. For instance, the *Burgers kernel* with initial data a multiple of the Dirac–delta function is

$$\theta(x,t;c,\varepsilon) = \sqrt{\frac{\varepsilon}{t}} \frac{(e^{c/2\varepsilon}-1)e^{-x^2/4\varepsilon t}}{\sqrt{\pi} + (e^{c/2\varepsilon}-1)\int_{x/\sqrt{4\varepsilon t}}^{\infty} e^{-y^2} dy},$$

(2.2.4) $\theta(x,0) = c\delta(x).$

This expression can also be obtained directly from the Burgers equation by the use of scaling analysis.

The Burgers kernel represents the time-asymptotic state of the general solutions with the same integral

$$\int_{-\infty}^{\infty} u(x,0)dx = c.$$

This is so not only for Burgers solutions but for solutions of convex conservation laws. In the following theorem we assume, for simplicity, that the initial value decays sufficiently fast so as to obtain the optimal convergence rate. The algebraic decay rate with exponent 3/2 is chosen for consistency with the system for which the same exponent comes up automatically from the coupling of waves of different characteristic families; see section 2.3.

THEOREM 2.2.1. *Suppose that $u(x,t)$ is a solution of the convex conservation law, normalized by $f(0) = f'(0) = 0$, $f''(0) = 1$, with initial value*

$$u_t + f(u)_x = u_{xx},$$

$$u(x,0) = O(1)\delta(|x| + 1)^{-3/2}, \qquad \int_{-\infty}^{\infty} u(x,0) = c.$$

Then, for δ small, $u(x,t)$ approaches the Burgers kernel $\theta(x,t) = \theta(x,t;c,1)$:

(2.2.5) $u(x,t) = \theta(x,t+1) + O(1)\delta(x^2 + t + 1)^{-3/4}.$

Proof. The total mass c of the diffusion wave θ is chosen to equal the total mass of the initial value so that

$$\int_{-\infty}^{\infty} v(x,t)dx = 0, \qquad v(x,t) \equiv u(x,t) - \theta(x,t+1).$$

This and the hypothesis on the decay of the initial value yield

$$v(x,0) = O(1)\delta(x^2 + 1)^{-3/4}, \qquad w(x,0) = O(1)\delta(x^2 + 1)^{-1/4},$$

$$w(x,t) \equiv \int_{-\infty}^{x} v(y,t)dy.$$

We have from the conservation law for $u(x,t)$ and the Burgers equation for $\theta(x,t)$ that

$$v_t = v_{xx} + \left[-\theta v - \frac{v^2}{2} + O(1)(|v|^3 + |\theta|^3) \right]_x.$$

For simplicity, we assume that the conservation law is the Burgers equation and the above equation is somewhat simplified:

$$v_t = v_{xx} + \left[-\theta v - \frac{v^2}{2} \right]_x.$$

Denote by $G(x,t)$ the heat kernel; then we have Duhamel's principle

$$v(x,t) = \int_{-\infty}^{\infty} G(x-y,t)v(y,0)dy$$

$$+ \int_0^t \int_{-\infty}^{\infty} \left[-\theta v - \frac{v^2}{2}\right]_y (y,s)G(x-y,t-s)dyds,$$

$$G(x,t) \equiv \frac{1}{\sqrt{4\pi t}}e^{-\frac{x^2}{4t}}.$$

We will also use another expression for the above integrals obtained by applying integration by parts:

$$-\int G_y(x-y,t)\left[\int_{-\infty}^y v(z,0)dz\right]dy,$$

$$\int\int\left[\theta v + \frac{v^2}{2}\right](y,s)G_y(x-y,t-s)dyds.$$

We will use repeatedly the simple estimate of the following form:

$$G_x(x,t) = \frac{1}{\sqrt{4\pi t}}\frac{2x}{4t}e^{-\frac{x^2}{4t}} = O(1)\frac{1}{t}e^{-\frac{x^2}{Dt}}$$

for any constant $D > 4$. For brevity, we will carry out the estimates for $t > 1$. From the decay of the initial values the first integral in the above expression for $v(x,t)$ is, for $|x| < \sqrt{t}$,

$$\int_{-\infty}^{\infty} G(x-y,t)v(y,0)dy = -\int_{\infty}^{\infty} G_y(x-y,t)w(y,0)dy$$

$$= \int_{-\infty}^{\infty} O(1)t^{-1}e^{-\frac{(x-y)^2}{Dt}}\delta(1+y^2)^{-1/4}dy$$

$$= \int_{|y|>\sqrt{t}} O(1)\delta t^{-1}e^{-\frac{(x-y)^2}{Dt}}t^{-1/2}dy + \int_{|y|<\sqrt{t}} O(1)\delta t^{-1}\delta(y^2+1)^{-1/4}dy$$

$$= O(1)\delta t^{-3/4} = O(1)\delta(x^2+t)^{-3/4}.$$

For $|x| > \sqrt{t}$,

$$\int_{-\infty}^{\infty} G(x-y,t)v(y,0)dy = -\int_{|y|<|x|/2} G_y(x-y,t)w(y,0)dy$$

$$+G(x-|x|/2,t)w(|x|/2,0) - G(x+|x|/2,t)w(-|x|/2,0)$$

$$+ \int_{|y|>|x|/2} G(x-y,t)v(y,0)dy$$

$$= O(1)\delta\left[t^{-1}e^{-\frac{x^2}{16t}}\int_{|y|<|x|/2}(y^2+1)^{-1/4}dy + t^{-1/2}e^{-\frac{x^2}{16t}}(x^2+1)^{-1/4}\right.$$

$$\left.+(x^2+1)^{-3/4}\int_{|y|>|x|/2} G(x-y,t)dy\right]$$

$$= O(1)\delta[t^{-1}e^{-\frac{x^2}{16t}}x^{1/2} + t^{-1/2}e^{-\frac{x^2}{16t}}(x^2+1)^{-1/4} + (x^2+1)^{-3/4}]$$
$$= O(1)\delta(x^2+1)^{-3/4} = O(1)\delta(x^2+t)^{-3/4}.$$

The above breakup of the integrals is to take advantage of the higher decay rates of the source and the Green function in their respective regions. For instance, we have used the estimate that, for $y < |x|/2$,

$$G(x-y,t) = \frac{1}{\sqrt{4\pi t}}e^{-\frac{(x-y)^2}{4t}} = O(1)\frac{1}{\sqrt{t}}e^{-\frac{x^2}{16t}}.$$

The double integral involves the solution $v(y,s)$ itself. This is dealt with usually by iterations, checking inductively that hypothesis (2.2.5) holds for each step. In each step of the iteration one estimates the double integral with the function $v(y,s)$ satisfying the pointwise assumption (2.2.5). We now carry this out. For this, we will need another repeatedly used estimate related to the semigroup property of the heat kernel:

$$\int_0^t \int_{-\infty}^{\infty} e^{-\frac{y^2}{D(s+1)}}e^{-\frac{(x-y)^2}{D(t-s)}}\,dy = \int_{-\infty}^{\infty} O(1)e^{-\frac{(t+1)[y-(s+1)x/(t+1)]^2}{D(s+1)(t-s)}}e^{-\frac{x^2}{D(t+1)}}\,dy$$

$$= O(1)(s+1)^{1/2}(t-s)^{1/2}(t+1)^{-1/2}e^{-\frac{x^2}{D(t+1)}}.$$

The first part is, for any constant $D > 4$,

$$\int_0^t \int_{-\infty}^{\infty} (\theta v)(y,s)G_y(x-y,t-s)$$

$$= \int_0^t \int_{-\infty}^{\infty} O(1)\delta^2(s+1)^{-5/4}e^{-\frac{y^2}{4(s+1)}}(t-s)^{-1}e^{-\frac{(x-y)^2}{D(t-s)}}\,dyds$$

$$= O(1)\delta^2 \int_0^t (t+1)^{-1/2}(s+1)^{-3/4}(t-s)^{-1/2}e^{-\frac{x^2}{D(t+1)}}\,ds$$

$$= O(1)\delta^2 \left[\int_0^{t/2} O(1)(t+1)^{-1/2}(s+1)^{-3/4}t^{-1/2}\,ds \right.$$

$$\left. + \int_{t/2}^t (t+1)^{-5/4}(t-s)^{-1/2}\,ds \right] e^{-\frac{x^2}{D(t+1)}}$$

$$= O(1)\delta^2(t+1)^{-3/4}e^{-\frac{x^2}{D(t+1)}} = O(1)\delta^2(x^2+t+1)^{-3/4}.$$

For the second part, we have

$$\int_0^t \int_{-\infty}^{\infty} (v^2)_y(y,s)G(x-y,t-s)\,dyds$$

$$= \int_0^t \int_{-\infty}^{\infty} O(1)\delta^2(y^2+s+1)^{-3/2}G_y(x-y,t-s)\,dyds$$

$$= \int_0^t \int_{|y|<\sqrt{s+1}} O(1)\delta^2(s+1)^{-3/2}(t-s)^{-1}e^{-\frac{(x-y)^2}{D(t-s)}}\,dyds$$

$$+ \int_0^t \int_{|y|>\sqrt{s+1}} O(1)\delta^2(y^2+1)^{-3/2}(t-s)^{-1}e^{-\frac{(x-y)^2}{D(t-s)}}\,dyds \equiv i_1 + i_2.$$

For $0 \le |x| \le 2\sqrt{t+1}$,

$$
\begin{aligned}
i_1 &= \int_0^t \int_{|y|<\sqrt{s+1}} O(1)\delta^2(s+1)^{-3/2}(t-s)^{-1/2}(t-s+1)^{-1/2}ds \\
&= \int_0^t O(1)\delta^2(s+1)^{-1}(t-s)^{-1/2}(t-s+1)^{-1/2}ds \\
&= O(1)\delta^2(t+1)^{-1}\ln(t+1) = O(1)\delta^2(x^2+t+1)^{-3/4},
\end{aligned}
$$

$$
\begin{aligned}
i_2 &= \int_0^{t/2} \int_{|y|>\sqrt{s+1}} O(1)\delta^2(y^2+1)^{-3/2}(t-s)^{-1}dyds \\
&\quad + \int_{t/2}^t \int_{|y|>\sqrt{s+1}} O(1)\delta^2(s+1)^{-3/2}(t-s)^{-1}e^{-\frac{(x-y)^2}{D(t-s)}}dyds \\
&= \int_0^{t/2} O(1)\delta^2(s+1)^{-1}(t-s)^{-1}ds \\
&\quad + \int_{t/2}^t O(1)\delta^2(s+1)^{-3/2}(t-s)^{-1/2}ds \\
&= O(1)\delta(t+1)^{-1}\ln(t+1) = O(1)\delta^2(x^2+t+1)^{-3/4}.
\end{aligned}
$$

For $|x| \ge 2\sqrt{t+1}$,

$$
\begin{aligned}
i_1 &= \int_0^t \int_{|y|<\sqrt{s+1}} O(1)\delta^2(s+1)^{-3/2}(t-s)^{-1}e^{-\frac{x^2}{4Dt}}dyds \\
&= \int_0^t O(1)\delta^2(s+1)^{-1}(t-s)^{-1}e^{-\frac{x^2}{4Dt}}dyds \\
&= O(1)\delta^2(x^2+t+1)^{-3/2},
\end{aligned}
$$

$$
\begin{aligned}
i_2 &= \int_0^t \int_{|x|/2>|y|>\sqrt{s+1}} O(1)\delta^2(y^2+1)^{-3/2}(t-s)^{-1}e^{-\frac{x^2}{4D(t-s)}}dyds \\
&\quad + \int_0^t \int_{|y|>|x|/2} O(1)\delta^2(x^2+1)^{-3/2}(t-s)^{-1}e^{-\frac{(x-y)^2}{D(t-s)}}dyds \\
&= \int_0^t O(1)\delta^2(s+1)^{-1}(t-s)^{-1}e^{-\frac{x^2}{4Dt}}dyds \\
&\quad + \int_0^t O(1)\delta^2(x^2+1)^{-3/2}(t-s)^{-1/2}ds \\
&= O(1)\delta^2(x^2+t+1)^{-3/4}.
\end{aligned}
$$

This completes the proof of the theorem. $\square$

As an immediate consequence of the theorem, we have

$$|v|_{L_p(x)} = O(1)(t+1)^{-3/4+1/(2p)}.$$

This is not the optimal rate for a scalar equation. If we assume the initial value to decay at higher rate,

$$u(x,0) = O(1)\delta(|x|+1)^{-\alpha}, \ \alpha \geq 2,$$

then we will obtain

$$v(x,t) = O(1)\delta \left[(x^2+t+1)^{-\alpha/2} + (t+1)^{-1} e^{-\frac{(x+1)^2}{D(t+1)}} \right].$$

In this case the optimal decay rate is obtained as follows:

$$|v|_{L_p(x)} = O(1)(t+1)^{-1+1/(2p)}.$$

However, we will see in the next section that the rate for a general system is strictly smaller than the optimal rate due to nonlinear coupling waves of distinct characteristic families.

The traveling wave, the *viscous shock wave* for a viscous conservation law connecting two states u_-, u_+, can be found in explicit form for the Burgers equation. The jump condition becomes $s = (u_+ + u_-)/2$. The ODE (2.1.5) with $f(u) = u^2/2$ becomes

$$\phi'(\xi) = \frac{1}{2}(\phi(\xi) - u_-)(\phi(\xi) - u_+).$$

This is integrated to yield

(2.2.6)
$$\phi(x,t) = \phi(x-st) = u_- + \frac{u_+ - u_-}{1 + e^{(u_+ - u_-)\xi/2}},$$
$$\xi \equiv \frac{x-st}{\varepsilon}, \ \phi(\pm\infty) = u_\pm.$$

The *viscous rarefaction wave* connecting two states can be found by the Hopf–Cole transformation. Here, for simplicity, we consider the waves that are symmetric with respect to the origin:

(2.2.7)
$$\psi(x,0) = \begin{cases} -u_0, & x < 0, \\ u_0, & x > 0, \end{cases}$$

$$u_0 > 0,$$

$$\psi(x,t) \equiv u_0 \frac{\phi_+(x,t) - \phi_-(x,t)}{\phi_+(x,t) + \phi_-(x,t)},$$

$$\phi_+(x,t) \equiv \frac{1}{\sqrt{4\pi t}} e^{-u_0 x/2 + u_0^2 t/4} \int_0^\infty e^{-\frac{(x-u_0 t - \eta)^2}{4t}} d\eta,$$

(2.2.8)
$$\phi_-(x,t) \equiv \frac{1}{\sqrt{4\pi t}} e^{u_0 x/2 + u_0^2 t/4} \int_{-\infty}^0 e^{-\frac{(x+u_0 t - \eta)^2}{4t}} d\eta.$$

The effect of the interplay of nonlinearity and dissipation on the stability of Burgers nonlinear waves can be understood by studying the Green function for the Burgers equation linearized around these waves. This will be done together with the study of the pointwise stability of nonlinear waves for systems in the subsequent sections. For now we will consider the weaker version of the stability of these waves.

THEOREM 2.2.2. *Consider a small perturbation of the stationary Burgers shock* $(u_0, -u_0)$:

$$u(x,0) = \phi(x) + \bar{u}(x,0),$$
$$\bar{u}(x,0) = \delta(|x|+1)^{-\beta}, \qquad \beta > \frac{3}{2} + \varepsilon,$$

for small δ and positive ε. Then the solution of the Burgers equation approaches a translation x_0 of the shock:

$$\sup_x v(x,t) \to 0 \text{ as } t \to \infty,$$
$$v(x,t) \equiv u(x,t) - \phi(x + x_0 - st),$$
$$x_0 \equiv \frac{1}{u_+ - u_-} \int_{-\infty}^{\infty} u(x,0)dx.$$

Note that the translation x_0 of the shock induces a mass proportional to the difference of the end state:

$$\int_{-\infty}^{\infty} (\phi(x + x_0 - st) - \phi(x - st))dx = x_0(u_+ - u_-).$$

This is easily shown by checking that it holds for $x_0 = 0$ and that the differentiation of both sides with respect to x_0 equals $u_+ - u_-$. Thus the translation x_0 is defined so that the perturbation has zero mass:

$$\int_{-\infty}^{\infty} v(x,t)dx = 0, \qquad t \geq 0.$$

Proof. We use the energy method, which is always important for the study of nonlinear waves. It provides an easy proof of the nonlinear stability of viscous shocks for scalar conservation laws, although without the pointwise estimates. This is done as follows.

Note that $v(x,0) \in L_1(x)$ by the convenient choice of $\beta > 1$. The zero mass of the perturbation $v(x,t)$ is made use of by considering its antiderivative $w(x,t)$:

$$v_t + (\phi v)_x = v_{xx} - \frac{1}{2}(v^2)_x,$$
$$w_t + \phi w_x = w_{xx} - \frac{1}{2}(w_x)^2, \qquad w(x,t) \equiv \int_{-\infty}^{x} v(y,t)dy.$$

Multiply the last equation by w and integrate to yield

$$\int_{-\infty}^{\infty} \frac{1}{2}w^2(x,t_2)dx + \int_{t_1}^{t_2} \int_{-\infty}^{\infty} \left[w_x^2(1 + \frac{1}{2}w) + \frac{1}{2}|\phi_x|w^2\right] dxdt$$

$$= \int_{-\infty}^{\infty} \frac{1}{2} w^2(x, t_1) dx.$$

With the crucial compression property of the shock $\phi_x < 0$ and the a priori hypothesis that the perturbation w is small, the above provides the basic energy estimate

$$\int_{-\infty}^{\infty} w^2(x, t_2) dx + \int_{t_1}^{t_2} \int_{-\infty}^{\infty} [v^2 + |\phi_x| w^2](x, t) dx dt = O(1) \int_{-\infty}^{\infty} w^2(x, t_1) dx,$$
$$0 \le t_1 \le t_2.$$

Here we have noted that $w_x = v$. We obtain a similar energy estimate by integrating the equation for v times v and the derivative of the equation for x times v_x:

$$\int_{-\infty}^{\infty} [v^2 + v_x{}^2](x, t_2) dx + \int_{t_1}^{t_2} \int_{-\infty}^{\infty} v_{xx}{}^2 dx dt + \int_{t_1}^{t_2} \int_{-\infty}^{\infty} v_x{}^2 dx dt$$
$$= O(1) \int_{-\infty}^{\infty} (v^2 + v_x{}^2)(x, t_1) dx.$$

This is integrated from $t_1 = t_2 - 1$ to $t_1 = t_2$, and then we set $t = t_2$ to obtain

$$\int_{-\infty}^{\infty} v^2(x, t) dx = O(1) \int_{t-1}^{t} \int_{-\infty}^{\infty} (v^2 + v_x{}^2)(x, s) dx ds.$$

The above energy estimates yield

$$\int_{-\infty}^{\infty} [w^2 + v^2 + v_x{}^2](x, t_2) dx \quad + \quad \int_{t_1}^{t_2} \int_{-\infty}^{\infty} [v^2 + v_x{}^2 + v_{xx}{}^2] dx dt$$
$$= \quad O(1) \int_{-\infty}^{\infty} [w^2 + v^2 + v_x{}^2](x, t_1) dx.$$

From these we have

$$\int_{0}^{\infty} \int_{-\infty}^{\infty} (v^2 + v_x{}^2) dx ds < \infty,$$

and so

$$\int_{t}^{\infty} \int_{-\infty}^{\infty} (v^2 + v_x{}^2) dx ds \to 0$$

as t tends to infinity. We thus conclude that

$$\int_{-\infty}^{\infty} v^2(x, t) dx \to 0, \qquad t \to \infty,$$

and so

$$\frac{1}{2} v^2(x, t) = \int_{-\infty}^{x} v v_x(y, t) dy \le \left[\int_{-\infty}^{\infty} v^2(y, t) dy \int_{-\infty}^{\infty} v_x{}^2(y, t) dy \right]^{-1/2}$$
$$\to 0, \; t \to \infty.$$

This establishes the stability of the viscous shock in the $L_\infty(x)$-norm.

Notice that in the above estimate we require the initial value $w(x,0) \in L_2(x)$. For that we assume a higher rate of decay for the perturbation

$$v(x,0) = \delta(|x|+1)^{-3/2-\varepsilon}$$

for a constant $\varepsilon > 0$.

The energy method with a weighted norm is an effective method for the scalar equation to obtain the stability result with rates of convergence. The method is also effective in studying the stability of strong shocks for equations with non-convex flux. For systems to be discussed later, the exact method of pointwise estimates using Green functions has been fruitful. In fact, the energy method works for the system only for special initial data [7], [23].

We next consider the perturbation of the viscous rarefaction wave (2.2.7), (2.2.8): The fact that the rarefaction wave is diffusive allows us to use the Burgers rarefaction wave $\psi(x,t)$ to construct approximate rarefaction waves $U(x,t)$ for general scalar conservation laws $u_t + f(u)_x = u_{xx}$, $f''(u) > 0$, which is the same as with the study of the perturbation of the Burgers kernel. We set

$$f'(U(x,t)) \equiv \psi(x,t)$$

so that the Burgers equation for $\psi(x,t)$ yields

$$
\begin{aligned}
f''(U)(U_t + f(U)_x) &= \psi_t + f(\psi)_x = \psi_{xx} \\
&= (f''(U)U_x)_x = f''(U)U_{xx} + f^3(U)(U_x)^2.
\end{aligned}
$$

We then see that $U(x,t)$ is an approximate solution of the convex scalar conservation law

(2.2.9)
$$
\begin{aligned}
U_t + f(U)_x &= U_{xx} + E(x,t), \\
E(x,t) &\equiv \frac{f^{(3)}(U)}{f''(U)}(U_x)^2 = O(1)(U_x)^2.
\end{aligned}
$$

The error $E(x,t)$ decays faster than the other terms of the equation and so this is a good time-asymptotic approximation. Below we use the energy method to study the stability of the rarefaction wave. The pointwise stability will be considered in section 5. We will consider the perturbation of the rarefaction wave $U(x,1)$ at time $t = 1$ to avoid the initial singularity.

THEOREM 2.2.3. *Let $u(x,t)$ be the solution of the convex conservation law with initial value a perturbation of the rarefaction wave:*

$$
\begin{aligned}
u_t + f(u)_x &= u_{xx}, \qquad f''(u) > 0, \\
u(x,0) &= U(x,1) + O(1)\delta(|x|+1)^{-3/2};
\end{aligned}
$$

then $f'(u)(x,t)$ tends to the Burgers rarefaction wave $\psi(x,t)$:

$$\sup_x |v(x,t)| \to 0 \qquad \text{as } t \to \infty,$$
$$v(x,t) \equiv u(x,t) - U(x,t+1).$$

Proof. From (2.2.9) for $U(x,t)$ we have

$$v_t + \left(\psi v + \frac{1}{2} f''(U) v^2 \right)_x = v_{xx} + O(1)(\psi_x(x,t+1))^2 + (O(1)v^3)_x.$$

Multiply this by $v(x,t)$ and integrate to obtain

$$\int_{-\infty}^{\infty} \frac{1}{2} v^2(x,t_2) dx + \int_{t_1}^{t_2} \int_{-\infty}^{\infty} \left(\frac{1}{2} \psi_x v^2 + \frac{1}{6} \frac{f^{(3)}(U)}{f''(U)} \psi_x v^3 + (v_x)^2 \right) dxdt$$
$$= \int_{-\infty}^{\infty} \frac{1}{2} v^2(x,t_1) dx + \int_{t_1}^{t_2} \int_{-\infty}^{\infty} (O(1)v(\psi_x)^2 + O(1)v^3 v_x) dxdt.$$

Here we have used the following integrations by parts:

$$\int_{-\infty}^{\infty} (\psi v)_x v dx = - \int_{-\infty}^{\infty} \psi v v_x dx = - \int_{-\infty}^{\infty} \psi \left(\frac{v^2}{2} \right)_x dx = \int_{-\infty}^{\infty} \frac{1}{2} \psi_x v^2 dx,$$

$$\int_{-\infty}^{\infty} \frac{1}{2} (f''(U) v^2)_x v dx = - \int_{-\infty}^{\infty} \frac{1}{2} f''(U) v^2 v_x dx = \int_{-\infty}^{\infty} -\frac{1}{6} f''(U)(v^3)_x dx$$
$$= \int_{-\infty}^{\infty} \frac{1}{6} f^{(3)}(U) U_x v^3 dx = \int_{-\infty}^{\infty} \frac{1}{6} \frac{f^{(3)}(U)}{f''(U)} \psi_x v^3 dx.$$

By using the expansion of the rarefaction wave $\psi_x > 0$ and the a priori hypothesis that $\sup_x |v(x,t)| = \delta$ is small, we have from the above that

$$\int_{-\infty}^{\infty} \frac{1}{2} v^2(x,t_2) dx + \int_{t_1}^{t_2} \int_{-\infty}^{\infty} (|\psi_x| v^2 + (v_x)^2) dxdt$$
$$= O(1) \left[\int_{-\infty}^{\infty} v^2(x,t_1) dx + \int_{t_1}^{t_2} \int_{-\infty}^{\infty} (v(\psi_x)^2 + v^6) dxdt \right].$$

It follows from the explicit form of the Burgers rarefaction wave $\psi(x,t+1)$ that $\psi_x(x,t+1)$ decays at the rate $(t+1)^{-1}$ and with essential support of the width $t+1$, and so

$$\int_{t_1}^{t_2} \int_{-\infty}^{\infty} v(\psi_x(x,t+1))^2 dxdt$$
$$\leq \int_{t_1}^{t_2} \left(\int_{-\infty}^{\infty} v^2 dx \right)^{1/2} \left(\int_{-\infty}^{\infty} (\psi_x(x,t+1))^4 dx \right)^{1/2} dt$$
$$= O(1)\delta \int_{t_1}^{t_2} ((t+1)^{-3})^{1/2} dt = O(1)\delta(\sqrt{t_2+1} - \sqrt{t_1+1}),$$

where we have made another a priori assumption that $v(x,t)$ is small in the $L_2(x)$-norm:

$$\sup_t \int_{-\infty}^{\infty} v^2(x,t)dx = O(1)\delta^2.$$

From the Sobolev inequality

$$v^2(x,t) \leq 2 \left(\int_{-\infty}^{\infty} v^2(y,t)dy \cdot \int_{-\infty}^{\infty} v_y{}^2(y,t)dy \right)^{1/2}$$

we obtain

$$\int_{t_1}^{t_2} \int_{-\infty}^{\infty} v^6(x,t_2)dxdt$$

$$\leq \int_{t_1}^{t_2} \left(\sup_x v^4(x,t) \cdot \int_{-\infty}^{\infty} v^2(x,t)dx \right) dt$$

$$\leq \int_{t_1}^{t_2} \left(\left(\int_{-\infty}^{\infty} v^2(x,t)dx \cdot \int_{-\infty}^{\infty} (v_x)^2(x,t)dx \right) \cdot \int_{-\infty}^{\infty} v^2(x,t)dx \right) dt$$

$$= O(1)\delta^4 \int_{t_1}^{t_2} \int_{-\infty}^{\infty} (v_x)^2(x,t)dxdt.$$

We combine the above estimates to obtain the basic energy estimate

$$\int_{-\infty}^{\infty} \frac{1}{2}v^2(x,t_2)dx + \int_{t_1}^{t_2} \int_{-\infty}^{\infty} (|\psi_x|v^2 + (v_x)^2)dxdt$$

$$= O(1) \left[\int_{-\infty}^{\infty} \frac{1}{2}v^2(x,t_1)dx + \delta((t_2+1)^{-1/2} - (t_2+1)^{-1/2}) \right].$$

The above a priori assumptions are a consequence of the energy estimate for higher derivatives. The rest of the proof follows the standard procedure as in the proof of the last theorem; details are therefore omitted.

There is a basic difference between the rarefaction waves and the shock waves in terms of the effects of dissipation. Rarefaction waves are stable in $L_p(x)$-norm for any $p > 1$. However, they are not stable in the $L_1(x)$-norm. This can be seen by translating parts of the wave away from each other. The dissipation will smooth out the resulting nonsmoothness but will not bring them back together. This is a nonlinear hyperbolic effect, and the dissipation has little effect there. The shocks are compressive and the effect of dissipation is to have an exponential decaying tail effect against the compression. This is reflected also in the fact that it is possible and necessary to determine the time-asymptotic location of the shock through the conservation law while the opposite is true for the rarefaction wave.

The study of nonlinear waves for the Burgers equation will be useful for the study of general systems later. In the next section we start with the study of the weakly nonlinear behavior of a perturbation of a constant state. As we will see, this already gives rise to interesting and rich nonlinear wave interactions.

2.3 Diffusion Waves

We now turn to a system of viscous conservation laws, $\mathbf{u}(x,t) \in \mathbb{R}^n$, $n \geq 2$. The main point is to understand, beyond the phenomena already mentioned for scalar equations in the last section, the effects of the coupling of waves of different characteristic families. We consider in this section the simplest situation of dissipation of solutions about a constant state taken to be zero here. For simplicity, we consider the artificial viscosity

$$(2.3.1) \qquad \qquad \mathbf{u}_t + f(\mathbf{u})_x = \mathbf{u}_{xx}.$$

As will be seen later, the coupling of waves of different characteristic families gives rise to wave tails with an algebraic decaying exponent of 3/2. For this reason, and for convenience, we assume that the initial value has the same algebraic decay rate

$$(2.3.2) \qquad \qquad \mathbf{u}(x,0) = O(1)\delta(|x|+1)^{-3/2}.$$

We are interested in the weakly nonlinear phenomenon, which corresponds to δ being small. We further simplify the situation by assuming that the inviscid system is strictly hyperbolic:

$$f'(\mathbf{u})r_i(\mathbf{u}) = \lambda_i(\mathbf{u})r_i(\mathbf{u}), \qquad l_i(\mathbf{u})f'(\mathbf{u}) = \lambda_i(\mathbf{u})l_i(\mathbf{u}), \qquad i = 1, 2, \ldots, n,$$

$$(2.3.3) \qquad \qquad \lambda_1(\mathbf{u}) < \lambda_2(\mathbf{u}) < \cdots < \lambda_n(\mathbf{u}).$$

Since (2.3.1) is dissipative, one may view it as the perturbation of the linear system

$$\mathbf{u}_t + f'(0)\mathbf{u}_x = \mathbf{u}_{xx}$$

and conclude that the solution of (2.3.1), (2.3.2) tends to zero in $L_\infty(x)$ and $L_2(x)$ as t tends to infinity. This weakly nonlinear analysis has been applied also to systems with physical viscosity and to other physical models such as the Boltzmann equation. We are interested, however, in deeper aspects of nonlinearity, such as the time-asymptotic behavior in the $L_1(x)$-norm. From the conservation law

$$(2.3.4) \qquad \qquad \int_{-\infty}^{\infty} \mathbf{u}(x,t)dx = \int_{-\infty}^{\infty} \mathbf{u}(x,0)dx,$$

the solution does not approach zero in the $L_1(x)$-norm. To study this problem, the first step is to calculate the leading term of the time-asymptotic state. From the weakly nonlinear analysis, the solution should dissipate mainly along the characteristic directions. We thus expand the solution in the eigenvectors at the zero state:

$$(2.3.5) \qquad \qquad \mathbf{u}(x,t) \equiv \sum_{i=1}^{n} u_i(x,)r_i, \qquad r_i = r_i(0).$$

Plug this into (2.3.1), expand, and diagonalize to yield

$$u_{it} + \lambda_i u_{ix} + \sum_{j,k=1}^{n} \left(\frac{1}{2} \eta_{ijk} u_j u_k \right)_x + (O(1)|\mathbf{u}|^3)_x = u_{ixx},$$

(2.3.6) $$\eta_{ijk} \equiv l_i f''(0) r_j r_k, \qquad \lambda_i = \lambda(0).$$

The leading term in the time-asymptotic expansion is obtained by ignoring the third-order term as well as the *off-diagonal* terms above:

(2.3.7) $$\theta_{it} + \lambda_i \theta_{ix} + \left(\frac{1}{2} \eta_{iii} \theta_i^2 \right)_x = \theta_{xx}.$$

This scalar equation is the linear heat equation with constant convection if $\eta_{iii} = 0$; otherwise, it is the Burgers equation.

The values η_{ijk} measure various aspects of the nonlinearity of the flux function $f(u)$. The particular values η_{iii} are the genuine nonlinearity identified by [11] mentioned in Chapter 1. To see this, we differentiate $f'(\mathbf{u}) r_k(\mathbf{u}) = \lambda_k(\mathbf{u}) r_k(\mathbf{u})$ in the direction of $r_j(\mathbf{u})$ to obtain

$$f''(\mathbf{u}) r_j(\mathbf{u}) r_k(\mathbf{u}) = (f'(\mathbf{u}) - \lambda_k(\mathbf{u})) r_k'(r_j(\mathbf{u})) + \lambda_k'(r_j(\mathbf{u})) r_k(\mathbf{u}).$$

Here $\lambda_k'(r_k(\mathbf{u}))$ denotes the differentiation of λ_i in the direction of r_k, etc. We then form the inner product of this with $l_i(\mathbf{u})$ to obtain

$$\eta_{ijk} = (\lambda_i(\mathbf{u}) - \lambda_k(\mathbf{u})) l_i(\mathbf{u}) \cdot r_k(r_j(\mathbf{u})) + \lambda_k'(r_k(\mathbf{u})) \delta_{ik}.$$

In particular, we have

(2.3.8) $$\eta_{iii} = \nabla \lambda_i(\mathbf{u}) \cdot r_i(\mathbf{u}), \qquad \mathbf{u} = 0.$$

Thus (2.3.7) is the heat equation (Burgers equation) if λ_i is linearly degenerate (genuinely nonlinear) at $\mathbf{u} = 0$; cf. (1.2.1), (1.2.6) of Chapter 1. We normalized r_i so that $\eta_{iii} = 1$ if the i-field is genuinely nonlinear:

(2.3.9) $$\begin{aligned} \theta_{it} + \lambda_i \theta_{ix} = \theta_{ixx}, & \qquad i\text{th field linearly degenerate,} \\ \theta_{it} + \lambda_i \theta_{ix} + (\tfrac{1}{2}\theta_i^2)_x = \theta_{ixx}, & \qquad i\text{th field genuinely nonlinear.} \end{aligned}$$

Thus η_{iii} represents the nonlinearity pertaining to a given characteristic family and is already present for the scalar equation. The off-diagonal terms with coefficients η_{ijk}, $(j,k) \neq (i,i)$, represent the nonlinearity of the coupling of waves of different characteristic families. This is the focus of our subsequent study in this section.

For given initial value (2.3.2) we decompose the conservation laws into

(2.3.10) $$\int_{-\infty}^{\infty} \mathbf{u}(x,0)dx \equiv \sum_{i=1}^{n} c_i r_i.$$

We are interested in solutions of (2.3.9) with integral c_i. Since the solutions of the heat and Burgers equations with given integral tend to the linear or nonlinear heat kernels, (2.2.4), we set

$$\theta_i(x,t) \equiv$$

(2.3.11)

$$
\begin{cases}
c_i \dfrac{1}{\sqrt{4\pi(t+1)}} e^{\frac{(x-\lambda_i(t+1))^2}{4(t+1)}}, & i\text{-field linearly degenerate,} \\[4mm]
\sqrt{\dfrac{1}{t+1}} \dfrac{(e^{c_i/2}-1)e^{-(x-\lambda_i(t+1))^2/4(t+1)}}{\sqrt{\pi}+(e^{c_i/2}-1)\int_{(x-\lambda_i(t+1))/\sqrt{4(t+1)}}^{\infty} e^{-v^2}\,dv}, & i\text{-field genuinely nonlinear.}
\end{cases}
$$

In (2.3.11) we have avoided the singularity at $t=0$ of the heat kernels by starting at time $t=1$. The approximate solution

$$\sum_{i=1}^{n} \theta_i(x,t) r_i$$

has the same integral as the solution $u(x,t)$. We therefore have

$$\mathbf{u}(x,t) \equiv \sum_{i=1}^{n} (\theta_i(x,t) + v_i(x,t)) r_i,$$

$$\int_{-\infty}^{\infty} v_i(x,t)\,dx = 0, \qquad i = 1,2,\ldots,n, \qquad t \geq 0,$$

(2.3.12)

$$v_{it} + \lambda_i v_{ix} = v_{ixx} + \left[-\frac{1}{2}\sum_{i\neq j}\eta_{jii}\theta_i^{\,2} - \sum_{(j,k)}\eta_{ijk}\theta_j v_k \right.$$

$$\left. \cdot\, O(1)(|\mathbf{v}|^2 + |\theta|^3) \right]_x.$$

The Green function for the linearized equation

$$v_t + \lambda v_x = v_{xx}$$

is the heat kernel

(2.3.13) $$G(x,t;y,s;\lambda) \equiv \frac{1}{\sqrt{4\pi(t-s)}} e^{-\frac{(x-y-\lambda(t-s))^2}{4(t-s)}}.$$

We study the solution of (2.3.12) by the Duhamel principle:

$$
\begin{aligned}
v_i(x,t) \;=\;& \int_{-\infty}^{\infty} G(x,t;y,0;\lambda_i) v_i(y,0)\,dy \\
(2.3.14)\qquad & + \int_0^t \int_{-\infty}^{\infty} G(x,t;y,s;\lambda_i) RHS_i(y,s)\,dy\,ds,
\end{aligned}
$$

$$RHS_i(x,t) \equiv \left[-\frac{1}{2}\sum_{i\neq j}\eta_{jii}\theta_i^{\,2} - \sum_{(j,k)}\eta_{ijk}\theta_j v_k \right.$$

$$+O(1)(|\mathbf{v}|^2 + |\theta|^3)\bigg]_x.$$

The usual procedure of *a priori estimates* involves the a priori assumption on the solution of (2.3.12) and then showing that, with $RHS_i(y,s)$ satisfying the assumption, the integral of (2.3.14) yields a function satisfying the same assumption. We state this assumption in the following theorem.

THEOREM 2.3.1. *Suppose that the initial value of* (2.3.1) *satisfies* (2.3.2), δ *small. Then the solution exists globally in time and satisfies* (2.3.12) *with the following decay rates:*

$$v_i(x,t) = O(1)[\delta(x - \lambda_i(0)(t+1))^2 + t + 1)]^{-3/4}$$

(2.3.15)
$$+ \sum_{j \neq i}((x - \lambda_j(0)(t+1))^3 + t^2 + 1)^{-1/2}].$$

Remark. The above estimate implies that

(2.3.16)
$$|\mathbf{v}|_{L_1(x)} = O(1)\delta(t+1)^{-1/4}$$

which is lower than the optimal rate $(t+1)^{-1/2}$ for the scalar equation. The rates cannot be improved even if the initial value $\mathbf{u}(x,0)$ is of compact support. This is due to the nonlinear coupling of waves of different characteristic families.

Proof. With assumption (2.3.2), identity (2.3.14), a priori hypothesis (2.3.15), and the integral property $\int v_i dx = 0$, we need only consider the following integrals: First, the contribution of the initial value in (2.3.14) is of the form

$$\mathrm{i} \equiv \int_{-\infty}^{\infty} \frac{1}{\sqrt{4\pi t}} e^{-\frac{(x-y-\lambda_i t)^2}{4t}} (O(1)(y^2+1)^{-1/4})_y dy.$$

This has been estimated in the proof of Theorem 2.2.1 for a scalar equation. The first double integrals in (2.3.14) are of the form

$$\mathrm{ii} \equiv \int_0^t \int_{-\infty}^{\infty} \frac{1}{\sqrt{4\pi(t-s)}} e^{-\frac{(x-y-\lambda_i(t-s))^2}{4(t-s)}} (\theta_j{}^2)_y(y,s)dyds, \qquad j \neq i,$$

$$\mathrm{iii} \equiv \int_0^t \int_{-\infty}^{\infty} \frac{1}{\sqrt{4\pi(t-s)}} e^{-\frac{(x-y-\lambda_i(t-s))^2}{4(t-s)}} (\theta_i v_i)_y(y,s)dyds.$$

From the a priori hypothesis (2.3.15) the remaining double integrals are of the form

$$\mathrm{iv} \equiv \int_0^t \int_{-\infty}^{\infty} \frac{1}{\sqrt{4\pi(t-s)}} e^{-\frac{(x-y-\lambda_i(t-s))^2}{4(t-s)}} (O(1)(s+1)^{-3/2} e^{-\frac{(y-\lambda_j(s+1))^2}{4(s+1)}})_y dyds,$$

$$\mathrm{v} \equiv \int_0^t \int_{-\infty}^{\infty} \frac{1}{\sqrt{4\pi(t-s)}} e^{-\frac{(x-y-\lambda_i(t-s))^2}{4(t-s)}}$$

$$\cdot (O(1)((y - \lambda_j(s+1))^2 + s + 1)^{-3/2})_y dy ds, \qquad i, j = 1, 2, \ldots, n.$$

In fact, iii should also contain the terms $\eta_{ijk}\theta_j v_k$, $j \neq k$. They are, however, absorbed into the terms ii and iv above because $v_k(x,t) = O(1)(t+1)^{-1}$ in the characteristic direction $x = \lambda_j t$. The term $\eta_{iii}\theta_i v_i$ needs to be treated separately and is listed as iii above. We need to show that the above integrals decay at rates no slower than those given by (2.3.15). The rate (2.3.15) is optimal. This is so because ii dominates and is of the same rate. The evaluation of these integrals involves long and tedious computations. It is this kind of computation that reveals the effects of nonlinear interactions of waves pertaining to different characteristic families. They are therefore essential for the understanding of the behavior of solutions for the system of viscous conservation laws. We will not carry out all the computations, but instead we make the following remarks of a qualitative nature. The integrals iv and v are estimated, taking advantage of the integral property $\int v dx = 0$ by first performing the integration by parts so that the integrands gain an additional factor of $(t - s)^{-1/2}$. One needs to break the integral into several parts according to the location of (x, t) relative to the characteristic lines $x - \lambda_j t$, $j = 1, 2, \ldots, n$. We illustrate this by calculating one of the simplest, telling terms: iv above with $i \neq j$. For simplicity we take $\lambda_i = 0$ and $\lambda_j = 1$:

$$\text{iv}_1 \equiv \int_0^t \int_{-\infty}^\infty \frac{1}{\sqrt{4\pi(t-s)}} e^{-\frac{(x-y)^2}{4(t-s)}} (O(1)(s+1)^{-3/2} e^{-\frac{(y-(s+1))^2}{4(s+1)}})_y dy ds$$

$$= \int_0^t O(1)(t-s)^{-1/2}(s+1)^{-1}(t+1)^{-1/2} e^{-\frac{(x-(s+1))^2}{D(t+1)}} ds.$$

Here we have applied the integration by parts before completing the square and integrating with respect to y. The constant D can be any constant greater than 4. For $|x| < \sqrt{t+1}$,

$$\text{iv}_1 = \int_0^t O(1)(t-s)^{-1/2}(s+1)^{-1}(t+1)^{-1/2}$$

$$= O(1)(t+1)^{-1}\ln(t+1) = O(1)(t+1)^{-1}\ln(t+1)e^{-\frac{x^2}{D(t+1)}}.$$

For $\sqrt{t+1} < x < t - \sqrt{t+1}$,

$$\text{iv}_1 = O(1)\left[\int_0^{x/M} (t+1)^{-1}(s+1)^{-1} e^{-\frac{x^2}{D(t+1)}} ds \right.$$

$$+ \int_{x/M}^{t-(t-x)/M} (t-x)^{-1/2}x^{-1}(t+1)^{-1/2} e^{-\frac{(x-(s+1))^2}{D(t+1)}} ds$$

$$\left. + \int_{t-(t-x)/M}^t (t-s)^{-1/2}(t+1)^{-3/2} e^{-\frac{(x-(t+1))^2}{D(t+1)}} ds \right]$$

$$= O(1)\left[(t+1)^{-1}(\ln x)e^{-\frac{x^2}{D(t+1)}} + (t-x)^{-1/2}x^{-1} + (t+1)^{-1}e^{-\frac{(x-(t+1))^2}{D(t+1)}} \right].$$

When the constant M is chosen to be large, D is close to 4. For the case $|x - t| < \sqrt{t+1}$,

$$\text{iv}_1 = O(1)\left[\int_0^{t/2}(t+1)^{-1/2}(s+1)^{-1}(t+1)^{-1/2}e^{-C(t+1)}ds\right.$$

$$+ \int_{t/2}^{t-\sqrt{t+1}}(t+1)^{-1/4}(t+1)^{-1}(t+1)^{-1/2}e^{-\frac{(x-(s+1))^2}{D(t+1)}}ds$$

$$\left.+ \int_{t-\sqrt{t+1}}^{t}(t-s)^{-1/2}(t+1)^{-1}(t+1)^{-1/2}ds\right]$$

$$= O(1)(t+1)^{-1} = O(1)(t+1)^{-1}e^{-\frac{(x-(t+1))^2}{D(t+1)}},$$

where C is a positive constant. For the case $x > t + \sqrt{t+1}$,

$$\text{iv}_1 = \int_0^t O(1)(t-s)^{-1/2}(s+1)^{-1}(t+1)^{-1/2}e^{-\frac{(x-(t+1))^2}{D(t+1)}}e^{-\frac{(t-s)^2}{D(t+1)}}ds$$

$$= O(1)(t+1)^{-1/2}e^{-\frac{(x-(t+1))^2}{D(t+1)}}\left[\int_0^{t/2}(t+1)^{-1/2}(s+1)^{-1}e^{-\frac{(t-s)^2}{D(t+1)}}ds\right.$$

$$+ \int_{t/2}^{t-\sqrt{t+1}}(t-s)^{-1/2}(t+1)^{-1}e^{-\frac{(t-s)^2}{D(t+1)}}ds$$

$$\left.+ \int_{t-\sqrt{t+1}}^{t}(t-s)^{-1/2}(t+1)^{-1}ds\right]$$

$$= O(1)(t+1)^{-1}e^{-\frac{(x-(t+1))^2}{D(t+1)}}.$$

A similar estimate holds for the case $x < -\sqrt{t+1}$. This completes the estimate of iv$_1$ and shows that it decays at the rate of (2.3.15).

The rationale for the complex breakup of the integral in estimating iv$_1$ can be understood in the following way: The diffusion waves propagate in the characteristic direction and diffuse with the width $\sqrt{s+1}$. The Green function, on the other hand, propagates in the negative time direction with width $\sqrt{t-s}$; see Figure 2.3.1.

This is the reason for the breakup of the integral with limits involving $\sqrt{t+1}$ in the above calculations. The estimates for the diffusion waves of algebraic type in v are more complicated, but they follow from the same principle.

The nonresonance condition $j \neq i$ for the term ii is important in the accuracy of $\sum_i \theta_i(x,t)r_i(0)$ as a time-asymptotic solution of (2.3.1). To make use of it we first notice that the functions $\theta_j{}^2(y,s)$ as defined in (2.3.11) dissipate along the characteristic direction λ_j and satisfy

(2.3.17) $\qquad (\theta_j{}^2)_s + \lambda_j(0)(\theta_j{}^2)_y = (\theta_j{}^2)_{yy} + O(1)\delta^2(s+1)^{-2}e^{-\frac{(x-\lambda_j(s+1))^2}{D(s+1)}}$

for any constant $D > 4$. The Green function

$$G_i(y,s) \equiv (4\pi(t-s))^{-1/2}e^{-\frac{(x-y-\lambda_i(t-s))^2}{4(t-s)}}$$

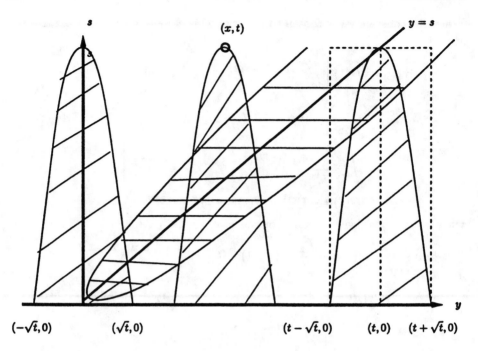

$(-\sqrt{t},0)$ $(\sqrt{t},0)$ $(t-\sqrt{t},0)$ $(t,0)$ $(t+\sqrt{t},0)$

Figure 2.3.1. Convolution of heat kernels.

satisfies

$$(2.3.18) \qquad G_{is} + \lambda_i G_{iy} + G_{iyy} = 0, \qquad G_i(x,t) = \delta(y-x).$$

With these estimates the term ii is dealt with in the following way: For small time, $0 < s < \sqrt{t+1}$, we take advantage of the faster decay rate of G_y; for larger time we take advantage of the faster decay rate of $(\theta_j{}^2)_y$:

$$\begin{aligned}
\text{ii} &= -\int_0^{\sqrt{t+1}} \int_{-\infty}^{\infty} G_{iy}(y,s)\theta_j{}^2(y,s)dyds \\
&\quad + \int_{\sqrt{t+1}}^t \int_{-\infty}^{\infty} G_i(y,s)(\theta_j{}^2)_y(y,s)dyds \equiv \text{ii}_1 + \text{ii}'.
\end{aligned}$$

To evaluate the above second integral ii', we need to make effective use of the cancelation of the convolution of the transversal Green function with the positive and negative parts of $(\theta_j{}^2)_y(y,s)$. For this we write

$$\begin{aligned}
(\theta_j{}^2)_y = \frac{1}{\lambda_i - \lambda_j}[((\theta_j{}^2)_s + \lambda_i(\theta_j{}^2)_y - (\theta_j{}^2)_{yy}) \\
-((\theta_j{}^2)_s + \lambda_j(\theta_j{}^2)_y - (\theta_j{}^2)_{yy})],
\end{aligned}$$

and so

$$\text{ii}' = \frac{1}{\lambda_i - \lambda_j} \left[\int_{\sqrt{t+1}}^{t} \int_{-\infty}^{\infty} +G_i(\theta_{j_s}^2 + \lambda_i \theta_{j_y}^2 - \theta_{j_{yy}}^2)(y,s)\,dy\,ds \right.$$

$$\left. + \int_{\sqrt{t+1}}^{t} \int_{-\infty}^{\infty} G_i(\theta_{j_s}^2 + \lambda_j \theta_{j_y}^2 - \theta_{j_{yy}}^2)(y,s)\,dy\,ds \right]$$

$$\equiv \text{ii}_2 + \text{ii}_3.$$

The integrand in ii_1 is the same as before, but the integration limit is only up to $\sqrt{t+1}$ and so the integral decays faster than it would if it were integrated up to the time t, e.g., the estimate for the term iv_1 above. Note, however, that, due to (2.3.17), the integrand in ii_3 has an additional decay rate of $(s+1)^{-1/2}$ as compared with that of the original form of ii. This again accounts for the satisfactory decay rate. The term ii_2 is treated using (2.3.18) and integration by parts:

$$\text{ii}_2 = -\frac{1}{\lambda_j - \lambda_i} \theta_j^2(x,t) + \int_{-\infty}^{\infty} G_i(y, \sqrt{t+1}) \frac{1}{\lambda_j - \lambda_i} \theta_j^2(y, \sqrt{t+1})\,dy,$$

which is of the form (2.3.15). It follows easily from (2.3.15) that the term iii is $O(1)\delta(t+1)^{-1/4}\theta_i(x,t)$, again of the form (2.3.15). We omit the straightforward details.

2.4 Viscous Shocks

Consider a viscous shock $\phi(x,t)$ for the system of viscous conservation laws with artificial viscosity

$$(2.4.1) \qquad\qquad \mathbf{u}_t + f(\mathbf{u})_x = \mathbf{u}_{xx}.$$

By a change of variable $x \to x - st$, $f(\mathbf{u}) \to f(\mathbf{u}) - s\mathbf{u}$, we may assume that the shock is stationary:

$$(2.4.2) \qquad\qquad f(\phi)_x = \phi_{xx}, \qquad \phi(\pm\infty) = \mathbf{u}_{\pm}.$$

The construction of shock profiles uses the central manifold theorem. Suppose that the pth characteristic field is genuinely nonlinear and that the shock is a p-shock so that we have the Lax entropy condition and the profile is compressive:

$$(2.4.3) \qquad\qquad \lambda_p(\mathbf{u}_-) > 0 > \lambda_p(\mathbf{u}_+), \qquad \frac{d}{dx}\lambda_p(\phi(x)) < 0.$$

In fact, there is a more precise description of weak shock profiles: From Theorem 1.2.1 we see that the shock speed is close to the arithmetic mean of the characteristics of its end states. Thus the shock propagation is governed approximately by the Burgers equation. The consideration of the shape of the shock profile is the same; the shock profiles are close to those of the Burgers shocks, (2.2.6). As a consequence, the Green function for the compression field

$$w_{pt} + \lambda_p(\phi(x))w_{px} = w_{pxx}$$

is close to that for the Burgers shock, (2.4.6). The compression field is therefore stable in the same way as for the Burgers shock. The stability analysis for the system also involves the study of waves pertaining to other characteristic fields transversal to the shock.

Before we study the viscous shocks for the system, we first complete the point-wise study for the Burgers shock. The study illustrates the strong stability effect of the compression. The first step is to study the Green function for the Burgers equation linearized around the shock wave $\phi(x)$, taken to be stationary: $u_- = -u_+ = u_0$, $u_\pm = \phi(\pm\infty)$. Consider the Burgers equation with unit viscosity $\varepsilon = 1$, linearized around the shock $\phi(x)$:

$$(2.4.4) \qquad\qquad v_t + (\phi v)_x = v_{xx}.$$

The Green function for this equation is easier to find if the equation is integrated:

$$(2.4.5) \qquad\qquad w_t + \phi w_x = w_{xx}.$$

Multiply this by

$$e^{-\frac{1}{2}\int^x \phi(y)dy}$$

and use the fact that ϕ is a stationary wave of the Burgers equation

$$\phi' = \frac{\phi^2}{2} - \frac{u_0{}^2}{2};$$

we have

$$z_t + \frac{1}{4}u_0{}^2 z = z_{xx}.$$

This can be transformed into a heat equation

$$(e^{\frac{u_0{}^2 t}{4}} z)_t = (e^{\frac{u_0{}^2 t}{4}} z)_{xx}.$$

Thus (2.4.5) has an explicit Green function as a weighted heat kernel

$$
\begin{aligned}
(2.4.6) \qquad \bar{G}(x,t;y,s) &= \frac{1}{\sqrt{4\pi(t-s)}} \frac{1+e^{u_0 y}}{1+e^{u_0 x}} e^{-\frac{(y-x+u_0(t-s))^2}{4(t-s)}} \\
&= \frac{1}{\sqrt{4\pi(t-s)}} \frac{1+e^{-u_0 y}}{1+e^{-u_0 x}} e^{-\frac{(y-x-u_0(t-s))^2}{4(t-s)}}.
\end{aligned}
$$

Here we have written two identical expressions of the Green function for later use. The Green function for (2.4.5) for $v(x,t)$ is

$$(2.4.7) \qquad\qquad G(x,t;y,s) = -\int_{-\infty}^{y} \bar{G}(x,t;\xi,s)_x d\xi.$$

Notice that from the explicit form of the Green function (2.4.6) that the propagation of information is along the characteristic direction $u_\pm$. However, it is weighted by the factor

$$\frac{1+e^{u_0 y}}{1+e^{u_0 x}} \qquad \text{or} \qquad \frac{1+e^{-u_0 y}}{1+e^{-u_0 x}}.$$

For instance, when $x > 0$ we may take the second expression in (2.4.6) and the Green function is similar to the linear heat kernel with speed $-u_0$. That is, it is similar to the inviscid case with straightforward superposition of diffusion. On the other hand, for the propagation of the source to the left, $y < 0$, we take the first expression in (2.4.6) and see that the propagation, although having speed u_0 as for the inviscid case, does not stop at the shock but continues to the right of the shock to reach (x, t). This diffusion effect diminishes as (x, t) moves away from the shock at the rate $e^{-u_0 x}$. The total effect has the consequence that the perturbation decays in time at the same rate as the decay rate of the initial perturbation at $x = \pm\infty$.

THEOREM 2.4.1. *Consider a small perturbation of the viscous Burgers shock:*

$$u(x,0) = \phi(x) + \bar{u}(x,0),$$
$$\bar{u}(x,0) = \delta(|x| + 1)^{-\beta}, \qquad \beta > 1,$$

for small δ. Then the solution of the Burgers equation approaches a translation of the shock:

$$v(x,t) \equiv u(x,t) - \phi(x + x_0 - st) = O(1)\delta(|x| + |u_+ - u_-|t + 1)^{-\beta},$$
$$x_0 = \frac{1}{u_+ - u_-} \int_{-\infty}^{\infty} \bar{u}(x,0)dx.$$

Proof. As in Theorem 2.2.2, the phase shift x_0 given above is determined through the conservation law

$$(2.4.8) \qquad \int_{-\infty}^{\infty} \bar{u}(x,t)dx = 0, \qquad t > 0.$$

As in the proof of Theorem 2.2.2, we have

$$w(x,t) \equiv \int_{-\infty}^{x} v(y,t)dy,$$
$$w(x,0) = O(1)\delta(|x| + 1)^{-\beta+1}.$$

From the Burgers equation this satisfies

$$w_t + \phi(x)w_x = w_{xx} - \frac{1}{2}w_x{}^2.$$

Instead of using the energy method, as in the proof of Theorem 2.2.2, we apply the Duhamel principle

$$w(x,t) = \int_{-\infty}^{\infty} w(x,0)G'(x,t;y,t)dy + \int_0^t \int_{-\infty}^{\infty} \frac{1}{2}w_x{}^2(y,s)G'(x,t;y,s)dyds.$$

For definiteness, set $x > 0$ and use the following combination of the expressions of the Green function in (2.4.6):

$$w(x,t) = \int_{-\infty}^{0} \frac{1}{\sqrt{4\pi t}} \frac{1 + e^{u_0 y}}{1 + e^{u_0 x}} e^{-\frac{(y-x+u_0 t)^2}{4t}} w(y,0)dy$$

$$+ \int_0^\infty \frac{1}{\sqrt{4\pi t}} \frac{1+e^{-u_0 y}}{1+e^{-u_0 x}} e^{-\frac{(y-x-u_0 t)^2}{4t}} w(y,0) dy$$

$$- \frac{1}{2} \int_0^t \int_{-\infty}^0 \frac{1}{\sqrt{4\pi(t-s)}} \frac{1+e^{u_0 y}}{1+e^{u_0 x}} e^{-\frac{(y-x+u_0(t-s))^2}{4(t-s)}} v^2(y,s) dy ds$$

$$- \frac{1}{2} \int_0^t \int_0^\infty \frac{1}{\sqrt{4\pi(t-s)}} \frac{1+e^{-u_0 y}}{1+e^{-u_0 x}} e^{-\frac{(y-x-u_0(t-s))^2}{4(t-s)}} v^2(y,s) dy ds.$$

As for the proof of Theorem 2.2.1, we use the usual iteration to evaluate these integrals. Thus the unknown functions $v(y,s)$ in the integrands are assumed to be of the form (2.4.8) so that the above integrals become

$$O(1)\delta \left[\int_{-\infty}^0 t^{-1/2} e^{-u_0 x} e^{-\frac{(y-x+u_0 t)^2}{4t}} (|y|+1)^{-\beta+1} dy \right.$$

$$+ \int_0^\infty t^{-1/2} e^{-\frac{(y-x-u_0 t)^2}{4t}} (|y|+1)^{-\beta+1} dy$$

$$+ \int_0^t \int_{-\infty}^0 (t-s)^{-1/2} e^{-u_0 x} e^{-\frac{(y-x+u_0(t-s))^2}{4(t-s)}} \delta(|y|+u_0 s + \sqrt{s}+1)^{-2\beta} dy ds$$

$$+ \left. \int_0^t \int_0^\infty (t-s)^{-1/2} e^{-\frac{(y-x-u_0(t-s))^2}{4(t-s)}} \delta(|y|+u_0 s + \sqrt{s}+1)^{-2\beta} dy ds \right]$$

$$\equiv O(1)[\text{i} + \text{ii} + \text{iii} + \text{iv}],$$

where we have assumed that $1 < \beta < 2$ for simplicity. We need to show that i and ii are of the same form as (2.4.8) and that iii and iv are smaller than (2.4.8). Clearly, ii dominates i and iv dominates iii. As before, we will deal with the case $1 < \beta < 2$ only. We now evaluate ii, first for the case $x + u_0 t > O(1)\sqrt{t}$:

$$\text{ii} = O(1)\delta \left[t^{-1/2} \int_0^{x+u_0 t + \sqrt{t}} (y+1)^{-\beta+1} dy \right.$$

$$+ \quad (x+u_0 t + \sqrt{t}+1)^{-\beta+1} \left. \int_{x+u_0 t+\sqrt{t}}^\infty t^{-1/2} e^{-\frac{(y-x-u_0 t)^2}{4t}} dy \right]$$

$$= \quad O(1)\delta(x+u_0 t)^{-\beta+1} = O(1)(x+u_0 t + \sqrt{t})^{-\beta+1}.$$

For the case $x + u_0 t < O(1)\sqrt{t}$,

$$\text{ii} = O(1)\delta \left[t^{-1/2} \int_0^{x+u_0 t + \sqrt{t}} (y+1)^{-\beta+1} dy \right.$$

$$+ \quad (x+u_0 t + \sqrt{t}+1)^{-\beta+1} \left. \int_{x+u_0 t+\sqrt{t}}^\infty t^{-1/2} e^{-\frac{(y-x-u_0 t)^2}{4t}} dy \right]$$

$$= \quad O(1)t^{(-\beta+1)/2} = O(1)(x+u_0 t + \sqrt{t})^{-\beta+1}.$$

The estimate for iv is similar and we treat only the case $x + u_0 t > O(1)\sqrt{t}$:

$$\text{ii} = O(1)\delta^2 \left[\int_0^t (t-s)^{-1/2} e^{-\frac{(y-x-u_0 t)^2}{Dt}} \int_{y+u_0 s < x+u_0 t} (y+u_0 s + \sqrt{s}+1)^{-2\beta} dy ds \right.$$

$$+ \int_0^t (t-s)^{-1/2} \int_{y+u_0 s > x+u_0 t} (y+u_0 s+1)^{-2\beta} dy ds \Big]$$

$$= O(1)\delta^2 \left[\int_0^t [(t-s)^{-1/2} e^{-\frac{(x+u_0 t)^2}{D(t-s)}} (u_0 + \sqrt{s}+1)^{-2\beta+1} ds \right.$$

$$+ \int_0^t (t-s)^{-1/2} (x+u_0 t+1)^{-2\beta+1} ds \Big]$$

$$= O(1)\delta^2 \left[t^{-\beta+1} e^{-\frac{(y-x-u_0 t)^2}{Dt}} + (t+1)^{1/2} x + u_0 t + 1)^{-2\beta+1} \right],$$

which is much smaller than the estimate for ii above.

This yields the same estimate for $w(x,t)$ as that for ii. The estimate for $v(x,t)$ stated in the theorem is done by using the differentiation of the integral representation for $w(x,t)$ above. The computation is similar to the above for $w(x,t)$; we omit the details. □

We note that the decay (2.4.8) is slow when the shock strength $2u_0$ is small. For weak shocks, information reaching the point (x,t) comes from sources almost directly below the point, while for strong shocks, the information comes from further out and thereby decays faster, at the rate of the decay of the initial value at $x = \pm\infty$.

We now turn to the stability of the viscous shocks for general systems. Consider a perturbation of the viscous profile

$$(2.4.9) \qquad \mathbf{u}(x,0) = \phi(x) + \bar{\mathbf{u}}(x,0), \qquad \bar{\mathbf{u}}(x,0) = \delta(|x|+1)^{-3/2}.$$

For the scalar conservation law, there is one degree of freedom, namely, the phase shift of the shock wave. As we have seen for the Burgers shocks, section 2.2, this is determined by the integral of the perturbation

$$\int_{-\infty}^\infty \bar{\mathbf{u}}(x,t)dx = \int_{-\infty}^\infty \bar{\mathbf{u}}(x,0)dx, \qquad t > 0.$$

For the system this yields n time invariants. In the meantime, the perturbation also gives rise to waves of other characteristic families $\lambda_i(\phi)$, $i \neq p$, leaving the shock and the phase shift of the shock. Waves leaving the shock diffuse around the constant state $\mathbf{u}_-$ (or $\mathbf{u}_+$) if it is slower (or faster) than the shock. These waves have been studied in the last section. We know that they carry an integral proportional to the right eigenvectors $r_i(\mathbf{u}_\pm)$. Denote the phase shift of the shock by x_0 and the diffusion waves by $\theta_i(x,t)r_i(\mathbf{u}_-)$, $i < p$; $\theta_i(x,t)r_i(\mathbf{u}_+)$, $i > p$, with weight c_i. Believing that the solution tends to such a time-asymptotic wave pattern

$$\mathbf{u}(x,t) \to \phi(x+x_0-st) + \sum_{i<p} \theta_i(x,t)r_i(\mathbf{u}_-) + \sum_{i>p} \theta_i(x,t)r_i(\mathbf{u}_+) \qquad \text{as } t \to \infty,$$

then we have that the conservation laws would yield

$$\int_{-\infty}^\infty \bar{\mathbf{u}}(x,0)dx = \int_{-\infty}^\infty \bar{\mathbf{u}}(x,t)dx = \int_{-\infty}^\infty (\mathbf{u}(x,t) - \phi(x-st))dx$$

$$\rightarrow \int_{-\infty}^{\infty} (\phi(x + x_0 - st) + \sum_{i<p} \theta_i(x,t) r_i(\mathbf{u}_-)$$

$$+ \sum_{i>p} \theta_i(x,t) r_i(\mathbf{u}_+) - \phi(x - st)) dx$$

$$= \int_{-\infty}^{\infty} (\phi(x + x_0 - st) - \phi(x - st)) dx + \sum_{i<p} c_i r_i(\mathbf{u}_-) + \sum_{i>p} c_i r_i(\mathbf{u}_+).$$

Here the diffusion waves are based on the end states $\mathbf{u}_\pm$:

(2.4.10)

$$\theta_i(x,t) \equiv \begin{cases} c_i \dfrac{1}{\sqrt{4\pi(t+1)}} e^{-\frac{(x-\lambda_i(\mathbf{u}_i^0)(t+1))^2}{4(t+1)}} , & i\text{th field linearly degenerate,} \\[2ex] \sqrt{\dfrac{1}{t+1}} \dfrac{(e^{c_i/2}-1)e^{-(x-\lambda_i(\mathbf{u}_i^0)(t+1))^2/4(t+1)}}{\sqrt{\pi} + (e^{c_i/2}-1)\int_{(x-\lambda_i(\mathbf{u}_i^0)(t+1))/\sqrt{4(t+1)}}^{\infty} e^{-y^2} dy} , \\ & i\text{th field genuinely nonlinear,} \end{cases}$$

$$\mathbf{u}_i^0 = \begin{cases} \mathbf{u}_-, & i<p, \\ \mathbf{u}_+, & i>p. \end{cases}$$

We therefore conclude that the phase shift x_0 of the shock and the strength c_i, $i \neq p$, of the diffusion waves that the perturbation $\bar{\mathbf{u}}(x.0)$ gives rise to are determined by

(2.4.11)
$$\int_{-\infty}^{\infty} \bar{\mathbf{u}}(x,0) dx = x_0(\mathbf{u}_+ - \mathbf{u}_-) + \sum_{i<p} c_i r_i(\mathbf{u}_-) + \sum_{i>p} c_i r_i(\mathbf{u}_+).$$

THEOREM 2.4.2. *Suppose that the perturbation $\bar{\mathbf{u}}(x,0)$ of the viscous shock $\phi(x-st)$ is small. Then the solution of the viscous conservation laws (2.3.1) exists globally in time and tends to the combination of shifted shocks and diffusion waves as follows:*

$$\mathbf{v}(x,t) \equiv \mathbf{u}(x,t) - \left[\phi(x + x_0) + \sum_{i\neq p} \theta_i(x,t) r_i(\mathbf{u}_0)\right] \equiv \sum_{i=1}^{n} v_i(x,t) r_i(\phi(x + x_0)),$$

$$v_i(x,t) = O(1)\delta \left[(x - \lambda_i(\mathbf{u}_0)(t+1)^2 + t + 1)^{-3/4} + \sum_{j\neq i,p} ((x - \lambda_j(\mathbf{u}_0)(t+1))^3 \right.$$

$$\left. + t^2 + 1)^{-1/2} + (|x| + 1)^{-1}(|x| + t + 1)^{-1/2}\right], \quad i \neq p,$$

$$v_p(x,t) = O(1)\delta \left[\sum_{j\neq p} ((x - \lambda_j(\mathbf{u}_0)(t+1))^3 + t^2 + 1)^{-1/2} \right.$$

$$\left. + (|x| + 1)^{-1}(|x| + t + 1)^{-1/2}\right].$$

Proof. As with the scalar shock, the location x_0 has been determined so that the residue $\mathbf{v}(x,t)$ has zero mass, which is made use of by considering the antiderivative $\mathbf{w}(x,t)$. The equations for $\mathbf{v}(x,t)$ are as follows:

$$\mathbf{v}_t + (f'(\phi)\mathbf{v})_x = \mathbf{v}_{xx} - [(f'(\phi+\theta) - f'(\phi))\mathbf{v}]_x$$

$$+ \left[O(1)(|\mathbf{v}|^2 + \theta^3) + \sum_{i,j \neq p,\ i \neq j} C_{ij}\theta_i{}^2 r_j(\mathbf{u}_i^0) \right]_x ,$$

$$\theta^2(x,t) \equiv \sum_{i \neq p} |\theta_i(x,t)|, \quad C_{ij} \equiv -\frac{1}{2} l_j f''(r_j, r_j)(\mathbf{u}_i^0).$$

We set

$$\mathbf{w}(x,t) \equiv \int_{-\infty}^{x} \mathbf{v}(y,t)dy,$$

$$\mathbf{w}(x,t) \equiv \sum_{i=1}^{n} w_i(x,t) r_i(\phi(x+x_0-st)), \quad v_i \equiv w_{ix},$$

$$\mathbf{v} = \sum_{i=1}^{n} v_i r_i(\phi) + \sum_{i=1}^{n} w_i r_i(\phi)_x.$$

From above we have

$$w_{it} + \lambda_i w_{ix} = w_{ixx} + \sum_{j \neq i,p} C_{ji}\theta_j{}^2 + \sum_{j \neq p} D_{ijk}\theta_j v_k + l_i$$

$$\cdot \left(\sum_{j=1}^{n} w_j \lambda_i r_{jx} + 2v_j r_{jx} + w_j r_{jxx} \right)$$

$$+ \sum_{j,k} E_{ijk} v_j v_k + O(1)(\theta^3 + |\mathbf{v}|^2 + |\mathbf{w}|^2 \phi_x{}^2) \text{ or}$$

$$w_{it} + \lambda_i w_{ix} = w_{ixx} + F_i(x,t),$$

$$v_{it} + (\lambda_i v_i)_x = v_{ixx} + \left[\sum_{j \neq i,p} C_{ji}\theta_j{}^2 + \sum_{j \neq p} D_{ijk}\theta_j v_k + l_i \right.$$

$$\cdot \left(\sum_{j=1}^{n} w_j \lambda_i r_{jx} + 2v_j r_{jx} + w_j r_{jxx} \right)$$

$$\left. + \sum_{j,k} E_{ijk} v_j v_k + O(1)(\theta^3 + |\mathbf{v}|^2 + |\mathbf{w}|^2 \phi_x{}^2) \right]_x ,$$

$$\lambda_i \equiv \lambda_i(\phi), \ l_i \equiv l_i(\phi), \ r_i \equiv r_i(\phi), \ v \equiv \sum_{i=1}^{n} |v_i|,$$

$$w^2 \equiv \sum_{i=1}^{n} |w_i|^2, \ \phi_x \equiv |\phi'|, \ D_{ijk} \equiv l_i f'' r_j r_k(\mathbf{u}_j^0).$$

Before performing the computation of the integrals from Duhamel's principle as we did for the diffusion waves in the last section, first we need to study the Green function for the linearized equations

$$w_{it} + \lambda_i(\phi(x))w_{ix} = w_{ixx}, \qquad i = 1, 2, \dots, n.$$

In the case of diffusion waves, the characteristic speed λ_i is evaluated at a constant state and the Green function is simply the heat kernel (2.4.10). We have variable convection speed $\lambda_i(\phi(x))$. For the compression characteristic field

$$\frac{d}{dx}\lambda_p(\phi(x)) < 0,$$

the Green function behaves like that for the scalar shock (2.4.6). For the transversal fields, $i \neq p$, the Green functions are close to the heat kernels with variable speeds. The study of these Green functions is a subject of interest because the traditional asymptotic method works only locally in time, while we need the accuracy for large time as well. That the Green functions exist is no problem. Systems with artificial viscosity can be diagonalized as above, and the Green function for the associated scalar equation can easily be estimated using the standard methods from functional analysis and the maximum principle for the scalar equation. We present here a construction of the approximate Green function $\rho(x, t; y, s)$ for the transversal field with positive speed $i > p$, $\lambda_i(\phi(x)) > 0$:

$$\rho_i(x, t; y, s) \equiv (4\pi(t - s))^{-1/2} e^{-\frac{[\lambda(y)(m(y) - m(x) + t - s)]^2}{4(t-s)}},$$

$$m'(y) \equiv \frac{1}{\lambda_i(\phi(y))}.$$

This satisfies the basic requirement of the Green function

$$\lim_{s \to t-} \rho(x, t; y, s) = \delta(y - x),$$

where δ is the Dirac–delta function. To solve the initial value problems

$$w_{it} + \lambda_i w_{ix} = w_{ixx} + F_i(x, t), \qquad w_i(x, 0) = w_{i0}(x),$$

we multiply the above equation, with (x, t) changing to the variables (y, s), by $\rho_i(x, t; y, s)$ and integrate over $-\infty < y < \infty, 0 < s < t$:

$$\begin{aligned}
w_i(x, t) &= \int_{-\infty}^{\infty} \rho_i(x, t; y, 0)w_0(y)dy + \int_0^t \int_{-\infty}^{\infty} \rho_i(x, t; y, s)F_i(y, s)dyds \\
&\quad + \int_0^t \int_{-\infty}^{\infty} [\rho_{is} + (\lambda_i(y)\rho_i)_y + \rho_{iyy}]w_i(y, s)dyds.
\end{aligned}$$

For ρ_i to be an accurate approximate Green function, we need to minimize the expression $\rho_{is} + (\lambda_i(\phi(y))\rho)_{iy} + \rho_{iyy}$. The required accuracy of $\rho_i(x, t; y, s)$ depends on the desired estimates on the wave interactions. The function $\lambda(x) = \lambda_i(\phi(x))$

is positive and depends smoothly on the shock $\phi(x)$, and since $\phi(x)$ tends to $\mathbf{u}_\pm$ exponentially we have

$$|\phi(x) - \mathbf{u}_-| + |\lambda(x) - \lambda_i(\mathbf{u}_-)| = O(1)\varepsilon e^{-\varepsilon|x|} \qquad \text{for } x < 0,$$
$$|\phi(x) - \mathbf{u}_+| + |\lambda(x) - \lambda_i(\mathbf{u}_+)| = O(1)\varepsilon e^{-\varepsilon|x|} \qquad \text{for } x > 0,$$
$$\lambda'(x) = O(1)\varepsilon^2 e^{-\varepsilon|x|}.$$

From these we have the following measure of accuracy of $\rho = \rho_i(x, t; y, s)$ as the Green function:

$$\rho_s + (\lambda\rho)_y + \rho_{yy} = \rho\lambda' + \rho(t-s)^{-1/2}\left[-\frac{3\lambda'}{2\lambda}H + \frac{\lambda'}{2\lambda}H^3 + \frac{\lambda'^2}{4\lambda^2}H^4\right]$$
$$+\rho\frac{-2\lambda^2\lambda' - \lambda\lambda'' + \lambda'^2}{2\lambda^2},$$
$$H \equiv \lambda(m(y) - m(x) + t - s)(t-s)^{-1/2}, \qquad \lambda \equiv \lambda(y).$$

From the above the accuracy of $\phi = \phi_i(x, t; y, s)$ is

$$\rho_s + (\lambda\rho)_y + \rho_{yy} = O(1)\varepsilon^2 e^{-\varepsilon|y|}(1 + (t-s)^{-1/2})\bar{\rho}$$
$$\bar{\rho} \equiv (4\pi(t-s))^{-1/2}e^{-\frac{[\lambda(y)(m(y)-m(x)+t-s)]^2}{D(t-s)}}$$

for any positive constant $D > 4 + O(1)\varepsilon$. The approximate Green function $\rho_i(x, t; y, s)$ can be estimated in terms of the heat kernels convected with speeds $\lambda_\pm = \lambda_i(\mathbf{u}_\pm)$:

$$G_{i\pm}(y, s) = G_\pm(x, t; y, s)$$
$$\equiv (4\pi(t-s))^{-1/2}\left[e^{-\frac{[y-x+\lambda_\pm(t-s)+C]^2}{D(t-s)}} + e^{-\frac{[y-x+\lambda_\pm(t-s)-C]^2}{D(t-s)}}\right],$$
$$G_{i0}(y, s) \equiv (4\pi(t-s))^{-1/2}\left[e^{-\frac{[y-\lambda_- x/\lambda_+ + \lambda_-(t-s)+C]^2}{D(t-s)}} + e^{-\frac{[y-\lambda_- x/\lambda_+ + \lambda_-(t-s)-C]^2}{D(t-s)}}\right],$$

for some positive constants C and $D > 4 + O(1)\varepsilon$. This needs to be done for various regions, say, $t - 1 < s < t$ versus $0 < s < t - 1$ and $x > \lambda_+(t-s)$, $y < 0$, versus $0 < x < \lambda_+(t-s)$, $y < 0$, and so forth. With such estimates, we have the expression for the solution

$$w_i(x, t) = \int_\infty^\infty O(1)G_+(y, 0)w_{i0}(y)dy + \int_0^t\int_{-\infty}^\infty O(1)G_{i+}(y, s)F_i(y, s)dyds$$

$$+ \int_0^t\int_{-\infty}^\infty O(1)\varepsilon^2 e^{-\varepsilon|y|}(1 + (t-s)^{-1/2})G_{i+}(y, s)w_i(y, s)dyds \text{ for } x > \lambda_+ t,$$

$$w_i(x, t) = \int_{-\infty}^\infty O(1)G_{i-}(y, 0)w_{i0}(y)dy + \int_{t-x/\lambda_+}^t\int_{-\infty}^\infty O(1)G_{i+}(y, s)$$

$$\cdot F_i(y, s)dyds + \int_0^{t-x/\lambda_+}\int_{-\infty}^\infty O(1)G_{i-}(y, s)F_i(y, s)dyds$$

$$+ \int_{t-x/\lambda_+}^{t} \int_{-\infty}^{\infty} O(1)\varepsilon^2 e^{-\varepsilon|y|}(1+(t-s)^{-1/2})G_{i+}(y,s)w_i(y,s)dyds$$

$$+ \int_{0}^{t-x/\lambda_+} \int_{-\infty}^{\infty} O(1)\varepsilon^2 e^{-\varepsilon|y|}(1+(t-s)^{-1/2})G_{i0}(y,s)w_i(y,s)dyds$$

for $\lambda_+ t > x > 0$,

$$w_i(x,t) = \int_{-\infty}^{\infty} O(1)G_{i-}(y,0)w_{i0}(y) + \int_{0}^{t} \int_{-\infty}^{\infty} O(1)G_{i-}(y,s)F_i(y,s)dyds$$

$$+ \int_{0}^{t} \int_{-\infty}^{\infty} O(1)\varepsilon^2 e^{-\varepsilon|y|}(1+(t-s)^{-1/2})G_{i-}(y,s)w_i(y,s)dyds \text{ for } x < 0.$$

Note that the $L_\infty(x)$ decay rate for $\mathbf{v}(x,t)$ is only $t^{-1/2}$, which is slower than the rate of $t^{-3/4}$ for the diffusion waves. This is due to the terms with $\mathbf{w}(x,t)$ in the equations for $v_i(x,t)$. (Since the coefficients of $\mathbf{w}(x,t)$ in the equations depend on the derivative of the shock profile, such terms are not present for the perturbation of a constant state.) Since this slow decaying term is essentially of finite width, the $L_1(x)$ decay rate is still governed by similar terms as for the diffusion waves and is therefore still $t^{-1/4}$. For details, see [13]. □

2.5 Viscous Rarefaction Waves

We first study the pointwise estimate for the scalar rarefaction wave $\psi(x,t)$ (2.2.8). First we study the Burgers equation linearized around the wave $\psi(x,t)$ and its antiderivative:

(2.5.1) $$v_t \quad + \quad (\psi v)_x = v_{xx},$$

(2.5.2) $$w_t + \psi w_x \quad = \quad w_{xx}, \qquad w_x \equiv v.$$

We now apply a transformation analogous to that of Hopf–Cole by multiplying (2.5.2) by

$$a(x,t) \quad \equiv \quad e^{-1/2 \int_{-\infty}^{x} (\psi(\xi,t)+u_0)d\xi}$$

to obtain the following constant coefficients equation:

$$z_t - u_0 z_x = z_{xx}, \qquad z(x,t) \equiv w(x,t)a(x,t).$$

The Green function for this is

$$\frac{1}{\sqrt{4\pi t}} e^{-((x-y)-u_0(t-s))^2 4(t-s)}.$$

Thus we have from the above identities that the Green functions $\bar{G}(x,t;y,s)$ for (2.5.2) and $G(x,t;y,s)$ for (2.5.1) are

(2.5.3) $$\bar{G}(x,t;y,s) = \frac{a(x,t)}{a(y,s)} \frac{1}{\sqrt{4\pi t}} e^{-((x-y)-u_0(t-s))^2 4(t-s)},$$

(2.5.4) $$G(x,t;y,s) = \int_{-\infty}^{y} \bar{G}(x,t;\xi,s)_x d\xi.$$

With this, we now study the pointwise stability of the rarefaction wave for general scalar convex conservation laws.

THEOREM 2.5.1. *Let $u(x,t)$ be the solution of the convex conservation law with initial value a perturbation of the rarefaction wave $U(x,t)$, (2.2.9):*

$$u_t + f(u)_x = u_{xx}, \qquad f''(u) > 0,$$
$$f'(u)(x,0) = U(x,1) + \delta(|x| + 1)^{-3/2};$$

then $f'(u)(x,t)$ tends to the Burgers rarefaction wave $\psi(x,t)$:

$$v(x,t) = \begin{cases} \delta(||x| - u_0 t|)^{-3/2}, & |x| > u_0 t + \sqrt{t+1}, \\ \delta(||x| - u_0 t|^{-1/2}, & ||x| - u_0 t| \le \sqrt{t+1}, \\ \delta\frac{1}{\sqrt{t+1}}\left(\frac{1}{|x-u_0 t|} + \frac{1}{|x+u_0 t|}\right), & |x| < u_0 t - \sqrt{t+1}, \end{cases}$$

$$v(x,t) \equiv u(x,t) - U(x,t+1).$$

Proof. From (2.2.8), (2.2.9) for $U(x,t)$ we have

$$v_t + \left(\psi v + \frac{1}{2}f''(U)v^2\right)_x = v_{xx} + O(1)(\psi_x(x,t+1))^2 + (O(1)v^3)_x.$$

We apply the Duhamel principle

$$\begin{aligned} v(x,t) &= \int_{-\infty}^{\infty} v(x,0)G(x,t;y,t)dy + \int_0^t \int_{-\infty}^{\infty} O(1)(\psi_x(y,s+1))^2 \\ &\quad + (O(1)v^3(y,s))_x G(x,t;y,s)dyds. \end{aligned}$$

Through lengthy, though straightforward, computations, we obtain pointwise estimates for the Green function in the regions outside, on the edges of, and inside the rarefaction waves. For instance, inside the rarefaction wave, the combination of the hyperbolic expansion and the heat dissipation gives rise to

$$\bar{G}(x,t;y,s) = O(1)\frac{1}{\sqrt{t(t-s)}}e^{-\frac{(y-xs/t)^2}{4s(t-s)}},$$
$$|x| < u_0 t - \sqrt{t}, \qquad |y| < u_0 s - \sqrt{t}.$$

The theorem is proved through the pointwise estimate of the above integrals. Details are omitted. □

For the Burgers equation, the pointwise estimate has been carried out using the Hopf–Cole transformation [8]. The study for systems is in [26].

2.6   Concluding Remarks

Hyperbolic and viscous conservation laws are the simplest models among a general class of dissipative physical systems. Other models in the class include the Euler equations for gas dynamics with thermononequilibrium, viscoelasticity with memory, the Boltzmann equation, and various kinetic models. There has been much progress in recent years in the field of conservation laws. Nevertheless, many, if not most, of the fundamental issues remain unresolved. These include, for instance, the problem of bridging the theory for a hyperbolic system and for the viscous system and the generalization of the theories to other physical models such as the Boltzmann equation and the interacting particle systems.

This book presents some of the basics of geometric shock wave theory. There is no attempt to give complete references here.

Bibliography

[1] A. Bressan, G. Goatin, and B. Piccoli, Well-posedness of the Cauchy problem for $n \times n$ systems of conservation laws, Mem. Amer. Math. Soc., to appear.

[2] R. Courant and K. O. Friedrichs, Supersonic Flow and Shock Waves, Interscience, 1948.

[3] C. M. Dafermos, Polygonal approximations of solutions of the initial value problem for a conservation law, J. Math. Anal. Appl., 38 (1972), 33–41.

[4] H. Freistuhler and T.-P. Liu, Nonlinear stability of overcompressive shock waves in a rotational invariant system of viscous conservation laws, Comm. Math. Phys., 153 (1993), 147–158.

[5] J. Glimm, Solutions in the large for nonlinear hyperbolic systems of equations, Comm. Pure Appl. Math., 18 (1965), 697–715.

[6] J. Glimm and P. Lax, Decay of Solutions of Systems of Hyperbolic Conservation laws, Mem. Amer. Math. Soc., AMS, Providence, RI, 1970.

[7] J. Goodman, Nonlinear asymptotic stability of viscous shock profiles for conservation laws, Arch. Rational Mech. Anal., 95 (1986), 325–344.

[8] Y. Hattori and K. Nishihara, A note on the stability of rarefaction waves for Burgers equation, Japan J. Indust. Appl. Math., 8 (1991), 85–96.

[9] K. Kawashima, Large-time behavior of solutions to hyperbolic-parabolic systems of conservation laws and applications, Proc. Roy. Soc. Edinburgh Sect. A, 106 (1987), 169–194.

[10] S. Kruskov, First-order quasilinear equations in several space variables, Mat. Sb., 123 (1970), 228–255 (in Russian); English translation in Math. USSR Sb., 10 (1970), 217–273.

[11] P. D. Lax, Hyperbolic systems of conservation laws II, Comm. Pure Appl. Math., 10 (1957), 537–566.

[12] P. D. Lax, Hyperbolic Systems of Conservation Laws and the Mathematical Theory of Shock Waves, CBMS Regional Conference Series in Applied Mathematics 11, SIAM, Philadelphia, 1973.

[13] T.-P. Liu, The deterministic version of the Glimm scheme, Comm. Math. Phys., 57 (1975), 135–148.

[14] T.-P. Liu, Admissible Solutions of Hyperbolic Conservation Laws, Mem. Amer. Math. Soc. 240, AMS, Providence, RI, 1980.

[15] T.-P. Liu, Linear and nonlinear large-time behaviour of solutions of hyperbolic conservation laws, Comm. Pure Appl. Math., 30 (1977), 767–796.

[16] T.-P. Liu, Nonlinear Stability of Shock Waves for Viscous Conservation Laws, Mem. Amer. Math. Soc. 308, AMS, Providence, RI, 1985.

[17] T.-P. Liu, Interaction of nonlinear hyperbolic waves, in Proceedings Nonlinear Analysis Conference. F.-C. Liu and T.-P. Liu, eds., 1989, Academia Sinica, Taipei, R.O.C.; World Scientific, River Edge, NJ, 171–184.

[18] T.-P. Liu, Pointwise convergence to shock waves for the system of viscous conservation laws, Comm. Pure Appl. Math., 50 (1997), 1113–1182.

[19] T.-P. Liu and T. Yang, L_1 stability for 2×2 systems of hyperbolic conservation laws, J. Amer. Math. Soc., 12 (1999), 729–774.

[20] T.-P. Liu and T. Yang, Well posedness of system of hyperbolic conservation laws, Comm. Pure Appl. Math., 52 (1999), to appear.

[21] T.-P. Liu and Y. Zeng, Large Time Behaviour of Solutions for General Quasilinear Hyperbolic-Parabolic Systems of Conservation Laws, Mem. Amer. Math. Soc. 599, AMS, Providence, RI, 1997.

[22] T.-P. Liu and K. Zumbrun, On nonlinear stability of general undercompressive viscous shock waves, Comm. Pure Appl. Math., 174 (1995), 319–345.

[23] A. Matsumura and K. Nishihara, On the stability of traveling wave solutions of a one-dimensional model system for compressible viscous gas, Japan J. Appl. Math., 2 (1985), 17–25.

[24] A. Majda and R. Pego, Stable viscosity matrices for systems of conservation laws, Trans. Amer. Math. Soc., 282 (1984), 749–763.

[25] J. Smoller, Shock Waves and Reaction-Diffusion Equations, Springer-Verlag, New York, 1982.

[26] A. Szepessy and K. Zumbrun, Stability of viscous conservation laws, Arch. Rational Mech. Anal., 133 (1996), 249–298.

[27] G. B. Whitham, Linear and Nonlinear Waves, Wiley, New York, 1974.

Index

admissible, 3, 15
amount of cancelation, 27
amount of interaction, 26, 27
approximate Green function, 66, 67

Burgers equation, 41, 53, 61
Burgers kernel, 41

characteristics, 1
conservation laws, 1, 63
contact discontinuity, 13
coupling of waves, 42, 52

degree of interaction, 16
diffusion waves, 52

elementary i-wave, 13
elementary waves, 14, 15, 21
energy method, 47, 49
entropy condition, 3, 8, 13, 15, 39, 40
equidistributed, 23, 28
equidistributedness, 34

generalized characteristics, 5, 6, 35,
        36, 38
genuinely nonlinear, 10, 13, 53
Glimm functional, 24
Green function, 44, 54, 57–60, 68

Hopf–Cole transformation, 41
Hugoniot curve, 13
Hugoniot curves, 10, 11, 19
hyperbolic, 1, 10

linearly degenerate, 13, 53

Random choice method, 21
Rankine–Hugoniot, 2
rarefaction wave, 10
regularity property, 37
Riemann problem, 2, 8, 14–16, 35

shock waves, 2
simple waves, 8

viscosity criterion, 40
viscous conservation laws, 39
viscous rarefaction wave, 46
viscous shock, 64
viscous shock wave, 46

wave curves, 13
wave interactions, 16, 26
wave partition, 30
wave tracing, 29, 35
weak solutions, 2